Biologically Inspired Optimization Methods

WITPRESS

WIT Press publishes leading books in Science and Technology.
Visit our website for the current list of titles.
www.witpress.com

WITeLibrary

Home of the Transactions of the Wessex Institute, the WIT electronic-library provides the international scientific community with immediate and permanent access to individual papers presented at WIT conferences. Visit the WIT eLibrary at
http://library.witpress.com

Biologically Inspired Optimization Methods

An Introduction

M. Wahde
Chalmers University of Technology, Sweden

M. Wahde
Chalmers University of Technology, Sweden

Published by

WIT Press
Ashurst Lodge, Ashurst, Southampton, SO40 7AA, UK
Tel: 44 (0) 238 029 3223; Fax: 44 (0) 238 029 2853
E-Mail: witpress@witpress.com
http://www.witpress.com

For USA, Canada and Mexico

WIT Press
25 Bridge Street, Billerica, MA 01821, USA
Tel: 978 667 5841; Fax: 978 667 7582
E-Mail: infousa@witpress.com
http://www.witpress.com

British Library Cataloguing-in-Publication Data

A Catalogue record for this book is available
from the British Library

ISBN: 978-1-84564-148-1

Library of Congress Catalog Card Number: 2008924944

*The texts of the papers in this volume were set
individually by the authors or under their supervision.*

No responsibility is assumed by the Publisher, the Editors and Authors for any injury and/or damage to persons or property as a matter of products liability, negligence or otherwise, or from any use or operation of any methods, products, instructions or ideas contained in the material herein. The Publisher does not necessarily endorse the ideas held, or views expressed by the Editors or Authors of the material contained in its publications.

© WIT Press 2008
Reprinted 2009

Printed in Great Britain by the MPG Books Group, Bodmin and King's Lynn

All rights reserved. No part of this publication may be reproduced, stored in a retrieval system, or transmitted in any form or by any means, electronic, mechanical, photocopying, recording, or otherwise, without the prior written permission of the Publisher.

For my parents

Contents

Abbreviations   xi

Preface   xiii

Notation   xvii

Acknowledgements   xix

1 Introduction
- 1.1 The importance of optimization 1
- 1.2 Inspiration from biological phenomena 2
- 1.3 Optimization of a simple behaviour for an autonomous robot 5

2 Classical optimization
- 2.1 Introduction 9
  - 2.1.1 Local and global optima 9
  - 2.1.2 Objective functions 10
  - 2.1.3 Constraints 11
- 2.2 Taxonomy of optimization problems 11
- 2.3 Continuous optimization 12
  - 2.3.1 Properties of local optima 12
  - 2.3.2 Global optima of convex functions 14
    - 2.3.2.1 Convex sets and functions 14
    - 2.3.2.2 Optima of convex functions 16
- 2.4 Algorithms for continuous optimization 16
  - 2.4.1 Unconstrained optimization 17
    - 2.4.1.1 Line search 17
    - 2.4.1.2 Gradient descent 19
    - 2.4.1.3 Newton's method 21
  - 2.4.2 Constrained optimization 24
    - 2.4.2.1 The method of Lagrange multipliers 25
    - 2.4.2.2 An analytical method for optimization under inequality constraints 29
    - 2.4.2.3 Penalty methods 30

2.5 Limitations of classical optimization 33
Exercises... 34

3 Evolutionary algorithms

3.1 Biological background 35
3.2 Genetic algorithms 40
    3.2.1 Components of genetic algorithms 46
        3.2.1.1 Encoding schemes 46
        3.2.1.2 Selection 48
        3.2.1.3 Crossover 52
        3.2.1.4 Mutation 53
        3.2.1.5 Replacement 55
        3.2.1.6 Elitism 55
        3.2.1.7 A standard genetic algorithm 55
        3.2.1.8 Parameter selection 56
    3.2.2 Properties of genetic algorithms 59
        3.2.2.1 The schema theorem 59
        3.2.2.2 Exact models 60
        3.2.2.3 Premature convergence 67
3.3 Linear genetic programming 72
    3.3.1 Registers and instructions 73
    3.3.2 LGP chromosomes 74
    3.3.3 Evolutionary operators in LGP 75
3.4 Interactive evolutionary computation 78
3.5 Biological vs. artificial evolution 82
3.6 Applications 83
    3.6.1 Optimization of truck braking systems 83
    3.6.2 Determination of orbits of interacting galaxies 86
    3.6.3 Prediction of cancer survival 92
Exercises... 96

4 Ant colony optimization

4.1 Biological background 100
4.2 Ant algorithms 104
    4.2.1 Ant system 105
    4.2.2 Max–min ant system 109
4.3 Applications 111
    4.3.1 Single-machine scheduling 112
    4.3.2 Co-operative transport using autonomous robots 114
Exercises... 116

5 Particle swarm optimization

5.1 Biological background 117
    5.1.1 A model of swarming 118

	5.2	Algorithm			120
	5.3	Properties of PSO			124
		5.3.1	Best-in-current-swarm vs. best-ever		125
		5.3.2	Neighbourhood topologies		125
		5.3.3	Maintaining coherence		126
		5.3.4	Inertia weight		127
		5.3.5	Craziness operator		128
	5.4	Discrete versions			129
		5.4.1	Variable truncation		129
		5.4.2	Binary PSO		130
	5.5	Applications			130
		5.5.1	Optimization of neural networks		131
			5.5.1.1	Prediction of pollutant levels	133
			5.5.1.2	Prediction of elephant migration patterns	134
		5.5.2	Optimization of cancer chemotherapy		136
	Exercises				137

6 Performance comparison

	6.1	Unconstrained function optimization	140
	6.2	Constrained function optimization	143
	6.3	Optimization of feedforward neural networks	145
	6.4	The travelling salesman problem	146

A Neural networks

	A.1	Biological background			151
		A.1.1	Neurons and synapses		151
		A.1.2	Biological neural networks		152
		A.1.3	Learning		153
			A.1.3.1	Hebbian learning	154
			A.1.3.2	Habituation and sensitization	154
	A.2	Artificial neural networks			156
		A.2.1	Artificial neurons		158
		A.2.2	Feedforward neural networks and backpropagation		159
			A.2.2.1	The Delta rule	159
			A.2.2.2	Limitations of single-layer networks	161
			A.2.2.3	Backpropagation	161
		A.2.3	Recurrent neural networks		169
		A.2.4	Other networks		171
	A.3	Applications			172

B Analysis of optimization algorithms

	B.1	Classical optimization		173
		B.1.1	Global minima of convex functions	173
		B.1.2	Properties of the gradient	174

  B.2 Genetic algorithms 174
    B.2.1 The schema theorem 174
    B.2.2 The genetic algorithm as a Markov process 176
      B.2.2.1 Number of populations of a given size 176
    B.2.3 Infinite population models 177
      B.2.3.1 Representing the crossover operator 177
      B.2.3.2 Initial distribution of chromosomes 178
      B.2.3.3 Elementary properties of binomial
           coefficients 178
      B.2.3.4 The mutation operator for functions of
           unitation 179
      B.2.3.5 Selection and mutations for the Onemax
           problem 180
    B.2.4 Expected runtime of a simple GA 181
    B.2.5 Estimating optimal mutation rates 182
  B.3 Ant colony optimization 183
    B.3.1 Pheromone limits in MMAS 183
    B.3.2 Convergence proof 184
    B.3.3 Runtime analysis for a simple ACO algorithm 184
  B.4 Particle swarm optimization 188
    B.4.1 Particle trajectories in PSO 188

C **Data analysis**

  C.1 Hypothesis evaluation 193
  C.2 Experiment design 200

D **Benchmark functions**

  D.1 The Goldstein–Price function 206
  D.2 The Rosenbrock function 206
  D.3 The Sine square function 207
  D.4 The Colville function 208
  D.5 A multidimensional benchmark function 208

Answers to selected exercises **209**

Bibliography **211**

Index **215**

Abbreviations

The abbreviations used throughout the book are compiled in the list below. Note that the abbreviations denote the singular form of the abbreviated words. Whenever the plural forms is needed, an *s* is added. Thus, for example, whereas GA abbreviates *genetic algorithm*, the abbreviation of *genetic algorithms* is written GAs. Some abbreviations that are only used once or twice, in a single section, have been left out from the list. Note also that an index of the technical terms introduced in the text can be found at the very end of the book.

ACO	Ant colony optimization
ANN	Artificial neural network
AS	Ant system
EA	Evolutionary algorithm
FFNN	Feedforward neural network
GA	Genetic algorithm
GP	Genetic programming
LGP	Linear genetic programming
NFL	No-free lunch (theorem)
MMAS	Max–min ant system
PSO	Particle swarm optimization
RNN	Recurrent neural network
TSP	Travelling salesman problem

Preface

The advent of rapid, reliable and cheap computing power over the last decades has transformed many, if not most, fields of science and engineering. The multidisciplinary field of optimization is no exception. First of all, with fast computers, researchers and engineers can apply classical optimization methods to problems of larger and larger size. In addition, however, researchers have developed a host of new optimization algorithms that operate in a rather different way than the classical ones, and that allow practitioners to attack optimization problems where the classical methods are either not applicable or simply too costly (in terms of time and other resources) to apply.

This book is intended as a course book for introductory courses in stochastic optimization algorithms,[1] and it has grown from a set of lectures notes used in courses, taught by the author, at the international master programme Complex Adaptive Systems at Chalmers University of Technology in Göteborg, Sweden. Thus, a suitable audience for this book are third- and fourth-year engineering students, with a background in engineering mathematics (analysis, algebra and probability theory) as well as some knowledge of computer programming.

The organization of the book is as follows: first, Chapter 1 gives an introduction to the topic of optimization. Chapter 2 provides a brief background on the important (and large) topic of classical optimization. Chapters 3–5 cover the main topics of the book, namely stochastic optimization algorithms inspired by biological systems. Three such algorithms, or rather classes of algorithms as there are many different versions of each type, are presented: Chapter 3 covers evolutionary algorithms, Chapter 4 ant colony optimization and Chapter 5 particle swarm optimization. In addition to a presentation of the biological background of the algorithms, each of these chapters contains examples and exercises. Chapter 6 contains a performance study, comparing the various algorithms on a set of benchmark problems, thus allowing the student to select appropriate parameter settings for specific problems and to assess the advantages and weaknesses of each method. The book has four appendices, covering neural networks (Appendix A), an analysis of (some of) the properties of optimization algorithms (Appendix B), a brief background on data analysis (Appendix C) and a list of benchmark functions (Appendix D). Demoting

---

[1] In this book, the terms *optimization method* and *optimization algorithm* will be used interchangeably.

the entire topic of neural networks to an appendix may, perhaps, seem a bit unorthodox. Why not place neural networks on the same footing as the other algorithms? Well, the main reason is that neural networks, *per se*, do not constitute an *algorithm* but rather a *computational structure* to which several algorithms can be applied. There are many optimization algorithms specifically intended for neural networks (such as backpropagation, described in Appendix A), but it is also possible to apply the algorithms presented in Chapters 3–5 in order to optimize a neural network. Thus, in this book, rather than taking centre stage, neural networks (of which there are *many* different kinds!) form a backdrop. Another reason is the fact that, at Chalmers University of Technology, and many other universities, neural networks are taught as a separate topic. Thus, the placement (in this book) of neural networks in an appendix should certainly not imply that the topic is unimportant, but rather that its importance is such that it merits its own course.

At Chalmers University of Technology, courses are taught in quarters lasting 7 or 8 weeks. Assuming an 8-week quarter, a suggested schedule for a course based on this book could be as follows. Week 1: Introduction, classical optimization methods (Chapters 1–2); Week 2–5: Evolutionary algorithms, neural networks and data analysis (Chapter 3 and Appendices A and C); Week 6: Ant colony optimization (Chapter 4); Week 7: Particle swarm optimization (Chapter 5); Week 8: Comparison of algorithms (Chapter 6 and Appendix D). The contents of Appendix B can be included along the way, but can also be skipped altogether or just briefly considered, should the course be geared towards applications rather than theory.

Clearly, with the 8-week constraint just mentioned, it is not feasible to cover *all* stochastic optimization methods; hence, those sampled in this book represent a subset of the available methods, and one that is hopefully not too biased. Optimization algorithms that have been left out include tabu search, simulated annealing and reinforcement learning (even though, in a general sense, all stochastic optimization algorithms can be considered as versions of reinforcement learning). In addition, related topics such as cellular automata, fuzzy logic, artificial life and so on are not covered either. Also, in the topics that are considered, it has been necessary to leave out certain aspects. This is so, since there exist numerous versions of the stochastic optimization algorithms presented in Chapters 3–5. Thus, for example, while Chapter 3 considers genetic algorithms, (linear) genetic programming and interactive evolutionary computation, related algorithms such as evolution strategies and evolutionary programming are not discussed. Similarly, only two versions of ant colony optimization are considered in Chapter 4. In general, the presentation is centered on practical applications of the various algorithms. More philosophical topics, such as complexity, emergence, the relation between biological and artificial life forms and so on will not be considered.

Furthermore, regarding applications, it should be noted that multi-objective optimization, that is, problems in which the objective function (see Chapter 2) is represented as a vector rather than a scalar, and where, consequently, the notion of optimality is generally replaced by so-called Pareto optimality, will not be considered. However, even though non-scalar objective functions are excluded from the presentation, this does not prevent us from considering simultaneous optimization with respect to several, possibly conflicting, objectives since, at least in some problems such as the single-machine weighted tardiness problem considered in Chapter 4, a scalar objective function can be formed as a weighted sum of the functions representing the individual objectives.

Despite the limitations, it is the author's hope that this book will provide the reader with a suitable background for pursuing further studies of stochastic (and other) optimization algorithms. As a guide to such endeavours, this book is concluded with a bibliography for further reading.

Notation

Z denotes the set of integers. **R** denotes the set of real numbers, and $\mathbf{R}^n$ its n-dimensional equivalent,

$$\mathbf{R}^n = \{(x_1, \ldots, x_n) : x_i \in \mathbf{R}, i = 1, \ldots, n\}. \tag{N1}$$

Similarly $\mathbf{Z}^n$ denotes the n-dimensional equivalent of **Z**. The notation [a,b] is used to denote a closed interval in **R**, i.e. the set $\{x : a \leq x \leq b\}$. Similarly,]a, b[denotes the open interval defined as $\{x : a < x < b\}$. As can be seen in eqn (N1), curly brackets are used for denoting sets in general. Curly brackets are also employed when listing a finite set of integers. For example $\{0, 1\}$ denotes the set consisting only of the two elements 0 and 1. The notation $A \subseteq B$ implies that A is a subset of B, meaning that every element of A is also an element of B. Vectors are written in bold lower-case characters, e.g. **x**. Here, **x** is to be understood as a column vector, i.e.

$$\mathbf{x} = \begin{pmatrix} x_1 \\ x_2 \\ \vdots \\ x_n \end{pmatrix}. \tag{N2}$$

To simplify the notation, however, a vector (or a point in $\mathbf{R}^n$) in component form is normally written $(x_1, x_2, \ldots, x_n)^T$, where T denotes the transpose. Note that some lists of variables that are *not* vectors, strictly speaking, are written without the transpose; the chromosomes introduced in Chapter 3, which are sometimes written $c = (g_1, \ldots, g_m)$, where g_i denotes gene i, constitute an example. $\|\mathbf{x}\|$ denotes the **Euclidean norm** of a vector $\mathbf{x} \in \mathbf{R}^n$, i.e.

$$\|\mathbf{x}\| = \sqrt{\sum_{i=1}^{n} x_i^2}. \tag{N3}$$

$\|\mathbf{x} - \mathbf{y}\|$ denotes the (Euclidean) distance between two points **x** and **y** in $\mathbf{R}^n$.

The variables appearing in optimization problems are normally written x_i, $i = 1, \ldots n$, where n denotes the number of variables. An exception occurs in Appendix A, where x_i is used for denoting the input elements in neural networks. However, the abuse of notation is slight, since the inputs often (but not always) represent the variables of the problem, for example in cases where a neural network is used to fit a function $f(x_1,\ldots,x_n)$. In Chapter 4, where the problems considered involve searching for paths in a graph rather than optimizing a mathematical function $f(x_1,\ldots,x_n)$, n instead denotes the number of nodes in the graph. Furthermore, n is used in different ways in the application examples concluding Chapters 3–5.

The stochastic optimization algorithms presented in Chapters 3–5 are all population-based, i.e. they maintain a set of candidate solutions to the problem at hand. The number of elements in this set (referred to as the population size for genetic algorithms, see Chapter 3, or the swarm size for particle swarm optimization, see Chapter 5) is denoted as N. Stochastic optimization is normally carried out by a computer program implementing the algorithm in question. In this book, the execution of such an algorithm will be referred to as a **run**.

The letters i, j, are typically reserved for integer counters. In some cases, variables contain both subscripts and superscripts. In those cases where there is a risk of confusion, superscripts are put in brackets in order to distinguish them from exponents. Thus, $c^{[j]}$ denotes a variable c with a superscript j, whereas c^j denotes the jth power of c. Some superscripts are, however, written without brackets, for example, x_{ij}^{pb} (Chapter 5), y_i^H (Appendix A), and $w_{ij}^{H \to O}$ (Appendix A). Since, throughout this book, exponents are always written using a single *lower-case* letter, there should be no risk of confusing the superscripts in the variables just listed with exponents.

Some algorithms involve iterations. In cases where there is sufficient space for a subscript or an argument (as in Chapter 2), the iterates are normally enumerated (e.g. as x_j or $x(j)$, whichever is most convenient for the application at hand). However, when, for example, a variable already has several subscripts, for instance, x_{ij}, new iterates are typically not enumerated explicitly. Instead, the next iterate is denoted as

$$x_{ij} \leftarrow x_{ij} + \Delta x_{ij}. \tag{N4}$$

A left-pointing arrow thus signifies that a new value is assigned to the variable shown to the left of the arrow. In addition, some elements of notation are only relevant to a particular chapter, and will therefore be described when introduced.

Whenever a new technical term is introduced and briefly described, it is written with **bold** letters. At the very end of the book, all technical terms are summarized in the form of an index.

Acknowledgements

I thank my family and friends, as well as colleagues and PhD students, for their patience and understanding during the writing of this book, and also for helping me with the proof-reading. I also express my gratitude to the many students who have suffered through early drafts of this book and have helped improve the book by finding misprints and other errors, most of which have hopefully been corrected. Any remaining errors are my own.

Furthermore, I would like to thank particularly K. S. Srikanth and the production team of Macmillan Publishing Solutions for their excellent typesetting. I am also grateful to Terri Barnett, Isabelle Strafford and Elizabeth Cherry at WIT Press, for their patience with my many delays. Last, but not least, I wish to thank Prof. Carlos Brebbia for inviting me to write this book in the first place.

Chapter 1

Introduction

1.1 The importance of optimization

Loosely speaking, **optimization** or **mathematical programming** as the topic is also known, refers to the process of manipulating a computational structure or system in order to achieve some pre-specified goal. As we shall see below, in optimization, the systems under consideration can often, but not always, be expressed in terms of a mathematical function, and the goal is then to find the minimum or maximum (depending on the application) of this function. Optimization plays a central role in science and engineering. In fact, there are so many examples of applications involving optimization that any list of such applications is bound to be incomplete and biased. Nevertheless, some classes of problems where optimization is highly relevant include scheduling, decision-making (e.g. financial portfolio optimization), microchip design (e.g. component placement on circuit boards), time-series prediction, path planning (e.g. in connection with autonomous robots), transportation-system design, various engineering problems (e.g. construction of bridges, vehicles etc.), and so forth.

In most practical applications of optimization, there are constraints, i.e. limits on the allowed range of the variables. In fact, the presence of constraints requires engineers to ask questions such as "Given that we have access to certain resources (such as construction material, time, financial resources etc.), what is the best we can do?" This is an informal definition of a constrained optimization problem.

Many problems can be formulated as the task of minimizing (or maximizing) a mathematical function, called the objective function, for the problem at hand. For such problems, all the tools of analysis, algebra, and other subdisciplines of mathematics can be summoned and, indeed, many optimization algorithms have been developed in mathematics. These algorithms define the field of classical optimization, and they are particularly well-developed for a class of optimization problems referred to as convex optimization problems, which will be defined in detail in the next chapter.

However, classical optimization algorithms also have shortcomings and they are not suitable for all optimization problems. Fortunately, there are plenty of alternative algorithms, most of which have been developed in the last few decades, and some of which are still under development. Even though the mathematical properties of such algorithms have been studied thoroughly (a process that is still very much ongoing), these algorithms have their origin not primarily in mathematics. Instead, they are inspired by biological phenomena. At a first glance, one may wonder why this is so. What, if anything, does biology have to do with optimization? One of the aims of this book is to answer this question, particularly in the first section of Chapters 3–5 in which the biological background of each algorithm is presented. However, it is perhaps appropriate to begin the book by a more general motivation for the use of biological inspiration in the development of optimization algorithms, and this will be our next topic.

1.2 Inspiration from biological phenomena

With the exception of the classical optimization algorithms that, by way of comparison, are presented in Chapter 2, the algorithms considered in this book (in Chapters 3–5) have two things in common: (1) they are stochastic and (2) they are inspired by biological phenomena. Informally expressed, the stochasticity of the algorithms imply that they rely, at least in part, on random numbers (even though this by no means implies that they are equivalent to a completely unguided, random search) and that different results *may* be obtained upon running such algorithms repeatedly.[1] By contrast, the classical optimization algorithms presented in Chapter 2 are **deterministic**, so that, starting from given initial conditions, the results obtained are always the same.

To practitioners of classical optimization methods, the stochasticity of the *results* obtained may seem a clear disadvantage, but there are, in fact, advantages as well. An example is the ability of stochastic optimization algorithms to find several different, but equally viable, solutions to a problem, a property that is useful in, for example, design optimization.

The second property, i.e. the inspiration taken from natural processes, may also require a motivation. Why would one base optimization algorithms on biological phenomena? One can readily note that nature is all about **adaptation**, which can be considered a form of optimization; the term adaptation refers to the gradual change

---

[1] The term **stochastic optimization** as it is used in this book, should not be confused with **stochastic programming**, which refers to methods for solving optimization problems involving stochastic objective functions. In this book, the optimization *algorithms* often contain stochastic operators, but the objective functions are deterministic in the applications considered.

Figure 1.1: *The skin of a shark, shown in the right panel, contains small, riblike structures lying in the direction of motion that allow the animal to swim faster. Left Panel: photo by the author. Right panel: photo by the Electron Microscope Unit, University of Cape Town. Reproduced with permission.*

in properties or behaviours in response to variations in the surroundings. Thus, for example, the growth of thick fur on arctic animals is an adaptation to cold weather; such animals are *optimized* for living under harsh conditions. Biological organisms display an amazing variety of adaptation.

Another excellent example, relevant for engineering optimization, is shark skin. Sharks have been around for a long time; in fact, fossil findings indicate that the ancestors of today's sharks appeared even before the dinosaurs. Thus, evolution, which forms the basis for the evolutionary algorithms considered in Chapter 3, has had a very long time to work on the design of these remarkable animals. If one touches a shark (which, perhaps, is not to be recommended), one can feel that its skin is not smooth. In fact, the skin of fast-swimming sharks is covered with small ribs (see Fig. 1.1) aligned with the direction of motion [2]. These structures, it turns out, affect the interaction between the skin of the shark and the tiny fluid vortices that appear as a result of movement. The net effect is a reduction in friction which enables the shark to swim faster. In fact, inspired by these findings, engineers are testing the effects of adding similar microstructures to the surfaces of ships and aircraft as well as the swimsuits of olympic swimmers. Obviously, sharks have not evolved their features in isolation: their prey has concurrently evolved various means of escaping, for example through coordinated swarm behaviour, a topic that will be considered further in Chapter 5. The simultaneous evolution of the properties of both predator and prey is an example of **co-evolution**, a phenomenon that has also been exploited in connection with stochastic optimization [30].

Another prime example is the evolution of the eye. As Dawkins [12] has noted, the eye has evolved in no less than 40 different ways, independent of each other. Two examples, shown in Fig. 1.2, are the compound eyes of insects and the lens eyes of mammals. Thus, evidently, there are many different solutions to the problem of generating a light-gathering device.

In addition to adaptation, many biological organisms also display *cooperation*. Even though cooperation is omnipresent in nature, nowhere is it as prevalent as in

4 BIOLOGICALLY INSPIRED OPTIMIZATION METHODS

 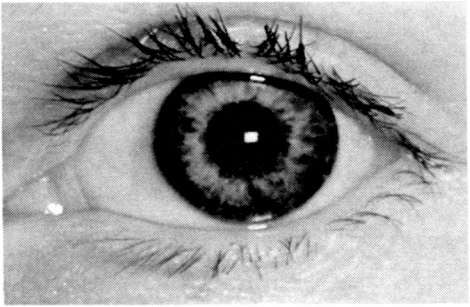

Figure 1.2: *Left panel: the compound eye of a fly. Right panel: a human eye. These are but two of the many light-gathering devices generated by biological evolution. Left Panel: micrograph courtesy of Greg Schmidt/Brian McIntyre, University of Rochester. Reproduced with permission. Right panel: photo by the author.*

certain species of insects, such as ants, bees, and termites. An example is the foraging behaviour of ants. Using only local communication, ants are able to coordinate their food gathering with amazing efficiency, and their behaviour forms the basis for ant colony optimization, studied in Chapter 4. Swarming behaviour, as seen in flocks of birds or schools of fish, is another form of cooperation which serves several purposes, such as improving the efficiency of food gathering and also protecting against predators. As indicated above, a school of fish, moving rapidly and largely unpredictably, presents a more confusing target for a marine predator than would a single fish. The behaviour of swarms has inspired particle swarm optimization, which will be considered in Chapter 5.

From the discussion above, it should be evident that there is ample motivation for basing optimization algorithms on biological phenomena. However, one should not exaggerate the analogy between the biological processes, on the one hand, and the optimization algorithms presented in Chapters 3–5, on the other. At most, those algorithms represent a mere caricature of their natural counterparts. This is all very well, since the intention, of course, is not to reproduce the biological processes, but rather to make use of those aspects that are relevant in the optimization problems occurring in science and engineering. We are thus free not only to distort or even discard the properties of biological phenomena that turn out not to be useful, but also to *add* properties that may not even exist in the biological phenomenon that serves as the inspiration.

However, before plunging into the field of optimization methods, we should say a few words about the so-called **No-free-lunch** (NFL) **theorem** [80]. The theorem concerns optimization algorithms based on search, of which the algorithms presented in Chapters 3–5 are specific examples. In such algorithms, candidate solutions to the problem at hand are generated and evaluated one after another. Essentially, the implication of the NFL theorem is that, *averaged over all possible problems* (or objective functions, see Section 2.1.2), no search algorithm outperforms any other algorithm. This would imply, for example, that a completely random search would do just as well as any of the stochastic optimization algorithms considered in this book. Yet, while the NFL theorem is certainly valid, it is rarely of practical significance, as one never considers *all* possible problems; in practice, it is found that, in a typical optimization problem, stochastic optimization algorithms such as genetic algorithms easily outperform random search. However, the theorem does imply that one should be careful before extrapolating the estimated performance of a stochastic optimization algorithm from one problem to another. Nevertheless, it is a fact that some algorithms *do* outperform others on specific classes of problems. For example, ant colony optimization, studied in Chapter 4, is particularly efficient when applied to problems involving path generation etc. In the final chapter of the book, we shall briefly return to the NFL theorem. Now, however, we will consider a motivational example illustrating stochastic optimization.

1.3 Optimization of a simple behaviour for an autonomous robot

In evolutionary robotics (ER) [53], which is a subfield of behaviour-based robotics, the aim is to generate robotic behaviours using artificial evolution, i.e. evolutionary optimization algorithms of the kind that will be presented in Chapter 3. In many cases, simulations are used in ER, as a precursor to the often costly procedure of constructing a real robot. Even though simulations serve an important role in robotics, the results obtained must eventually be transferred to real robots in order to be relevant. This is often a difficult procedure, involving several iterations of simulations and robot construction. Here, however, the transition from simulation to real robots will not be considered; instead, only a simple example will be studied, with the aim of introducing the topic of stochastic optimization. More specifically, the goal will be to generate a simple cleaning behaviour for a two-wheeled, differentially steered[2] robot such that, when executing this behaviour, the robot will clear an arena by moving garbage objects consisting of small cylinders out to the edges. The robot and the simulation setup are shown in Fig. 1.3. The equations of motion, which will not be given here, were derived from elementary kinematic and dynamic considerations [75]. The equations were then discretized, and integrated for T (simulated) seconds, with a time step of around 0.01 s. Each robot was equipped with five proximity sensors, modelled as infrared (IR) sensors. The sensors had a rather short range, about twice the diameter of the robot, and their readings s_i were normalized to the range [0, 1].

The movement of the robots was determined by simple control systems consisting of **if-then-else rules**. These control systems define a number of states, each with a given setting of the torques τ_L and τ_R applied to the two wheels. At initialization,

---

[2] A differentially steered robot is equipped with two independently driven wheels, one on each side of the body. For balance, such robots commonly also have two passive wheels, in the front and in the back.

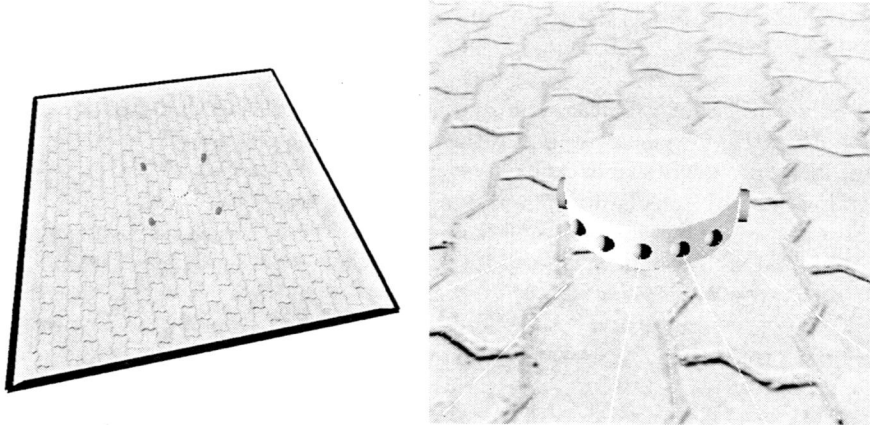

Figure 1.3: *Left panel: simulation setup, showing the robot and the four garbage objects. Right panel: a close-up of the robot. The rays emanating from its body indicate the opening angle and the range of the proximity sensors, shown as small spheres attached to the body.*

state one is active, determining the initial torque on the wheels. In every time step, a number of transition conditions (associated with the active state) are checked and, if a particular condition is satisfied, the robot jumps to a specified target state, which then becomes the active state. If no condition is satisfied, the robot remains in its current state.

The variables of the problem consist of the parameters used for the torque settings for each state, as well as the parameters determining the transition conditions. For the simulations presented here, it is hard to say, *a priori*, how many states are needed. Thus, in the simulations, the number of (used) states was allowed to vary, from 10 to 25. Each state required, for its definition, 21 parameters. Thus, the space of possible control systems, referred to as the search space (see Section 3.2), ranged in size from 210 to 525 dimensions.

Since the aim is to move the garbage objects to the edges of the arena, one may measure the performance of a given robot by considering the quantity F defined as

$$F = \sum_{j=1}^{n_o} \left(x_j^2(T) + y_j^2(T) \right), \tag{1.1}$$

where n_o denotes the number of garbage objects, and $(x_i(T), y_i(T))$ determines the location of object i at time T, i.e. at the end of the evaluation of a robotic control system. Clearly, a good control system will result in the robot reaching large values of F.

Now, this problem is quite different from a typical function optimization task of the kind that we will encounter in Chapter 2. In classical optimization, an objective function $f(x_1,\ldots,x_n)$, i.e. a quantity that should be minimized or maximized (depending on the problem) is normally given, and many classical optimization methods rely on the computation of gradients (essentially derivatives, see Section 2.3.1) of the objective function.

However, for the robot cleaning task, the value of the evaluation measure F is only obtained at the end of a rather lengthy simulation, in this case lasting 10,000 time steps, corresponding to $T = 100$ s. Thus, forming gradients of F would involve taking between 210 and 525 derivatives (one with respect to each variable) of F, a procedure that would require between $2 \times 210 = 420$ and $2 \times 525 = 1{,}050$ evaluations of F. This is where stochastic optimization algorithms come into play, by attempting to make better use of those function evaluations.

Furthermore, the variation in size of the control systems presents another difficulty for classical optimization methods, in which it is typically assumed that the number of variables is fixed and given at the outset. By contrast, some stochastic optimization algorithms are well suited for coping with systems of variable size. Here, a genetic algorithm was used, essentially of the kind described in Chapter 3, but with some modifications to handle the variation in the total number of variables described above. The details of the algorithm will not be given here. Instead, we shall briefly consider the results obtained.

In early generations, the simulated robots did little more than run around in circles near their starting position. However, some robots were lucky enough to hit one or a few garbage objects by chance, thereby moving them slightly towards the walls of the arena. Before long, there appeared robots that would hit *all* the garbage objects. The next evolutionary leap led to purposeful movement of objects. Here, an interesting method appeared: since both the body of the robot and the garbage objects were round, objects could not easily be moved forward; instead, they would slide away from the desired direction of motion. A method involving several if-then-else rules was found, in which the robot moved in a zig-zag fashion, thus managing to keep the garbage object in front. Next, robots appeared that were able to deliver a garbage object at a wall, and then return to the centre of the arena in order to detect remaining objects, moving backwards to avoid wasting time on an unnecessary turn. Towards the end of the run, the best-evolved robots were able to place three out of four garbage objects near a wall. A trajectory for such a robot is shown in Fig. 1.4. Had the simulation time been longer, the robot would probably have been able to move the fourth object as well.

We should note that the behaviour generated here might not generalize very well to other initial configurations of obstacles; the optimization algorithm is likely to exploit the exact initial placement of the obstacles as it improves the robotic behaviour. In order to obtain a more general behaviour, one would need to test each robot against several obstacle configurations, something that was not done in this case.

In addition to coping with variations in the number of variables and the absence of easily computed derivatives of the objective function, the algorithm generated somewhat different results in different runs, due to its stochastic nature. For example, several methods for moving garbage objects were found, some robots

8 BIOLOGICALLY INSPIRED OPTIMIZATION METHODS

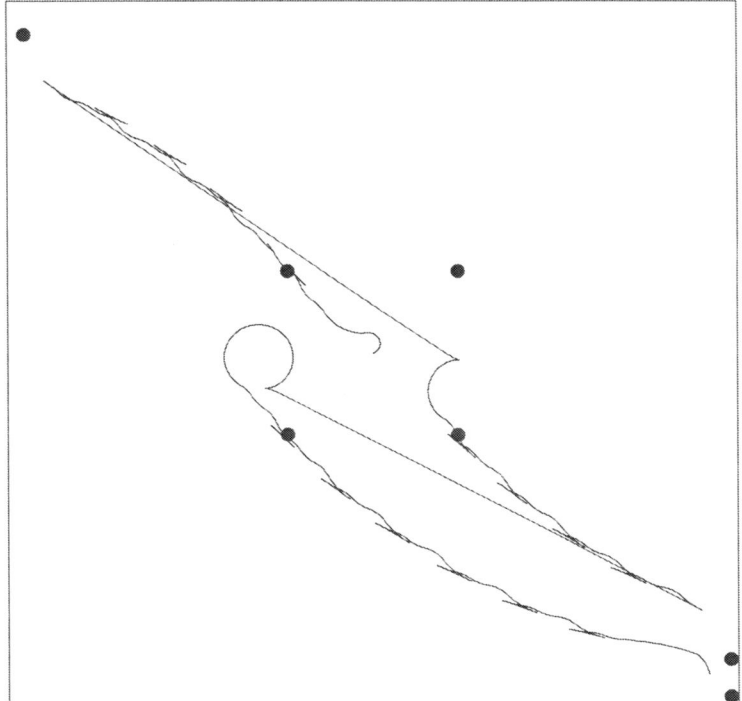

Figure 1.4: *The trajectory of one of the best robots obtained. The initial and final positions of the garbage objects are also shown. Note that one object was left untouched. The centre of the figure coincides with the origin of the coordinate system.*

favouring a slow, controlled motion whereas others used a more violent approach. Thus, the simulations also illustrated another property of stochastic optimization algorithms, namely their ability to find different solutions to a given problem. Before considering stochastic optimization methods in detail, however, we shall now first turn to classical optimization algorithms.

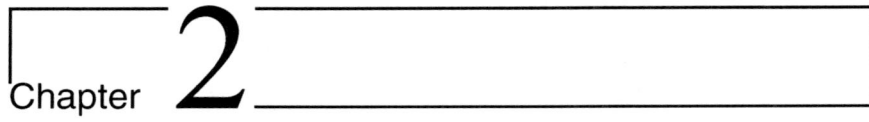

Chapter 2

Classical optimization

The topic of **classical optimization** [1], or mathematical programming as it is also known[1] has a long history, and has been attacked by, among many others, illustrious mathematicians such as Gauss and Newton. The topic comprises both analytical and numerical methods, and their generally deterministic nature distinguishes them from the stochastic optimization methods that form the core of this book. The aim of this chapter is to give a brief overview of classical optimization, before proceeding to the topic of stochastic optimization methods.

2.1 Introduction

2.1.1 Local and global optima

Let $I(x^*, \delta) = \{x: |x - x^*| < \delta\}$, where $\delta > 0$, denote a **neighbourhood** of x^*. To begin with, let us consider a function of a single variable x, i.e. $f: \mathbf{D} \to \mathbf{R}$, where $\mathbf{D} \subseteq \mathbf{R}$ is an open set. Such a function is said to have a **local minimum** at a point x^* if, in a neighbourhood of x^*, the function takes larger values than $f(x^*)$. Thus, more formally x^* is a (strict) local minimum of f if

$$\exists \delta > 0: f(x) > f(x^*) \quad \forall x \in I(x^*, \delta), \ x \neq x^*. \tag{2.1}$$

The concept of a **local maximum** is defined analogously. The terms **local optimum** and **local extremum** are used to denote either a local minimum or a local maximum. The definitions of local optima can be extended to functions $f: \mathbf{D} \to \mathbf{R}$,

---

[1] Even though some methods in mathematical programming depend on numerical computation, it should be noted that the term *mathematical programming* has no direct relation to the term *computer programming*.

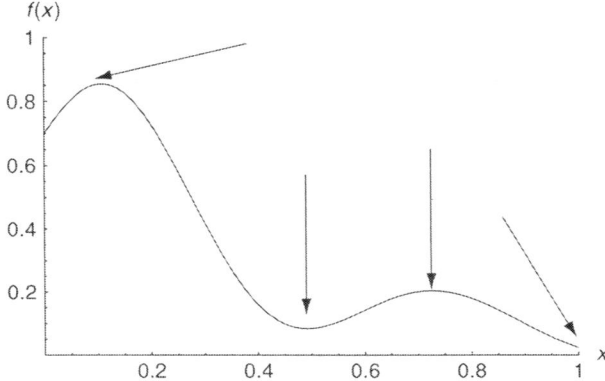

Figure 2.1: *The arrows indicate the location of local and global optima for a function defined on the closed interval* [0, 1]. *The leftmost arrow points to the global optimum, whereas the rightmost arrow indicates the global minimum. Note that the global minimum is not, strictly speaking, a local minimum, since a neighbourhood cannot be defined at* $x = 1$.

where $\mathbf{D} \subseteq \mathbf{R}^n$ again is an open set, by generalizing the concept of a neighbourhood to n dimensions, i.e. by considering points $\mathbf{x} = (x_1, x_2, \ldots x_n)$ in $\mathbf{R}^n$ such that $\|\mathbf{x} - \mathbf{x}^*\| < \delta$. A function $f(x)$ may have many local optima, as shown in Fig. 2.1. In addition to local optima, the concept of *global* optima is essential in optimization: a function $f: \mathbf{D} \to \mathbf{R}$, has a **global minimum** at a point x^* if $f(\mathbf{x}) \geq f(\mathbf{x}^*)\ \forall x \in \mathbf{D}$. The definition of a **global maximum** is analogous. Provided that $\mathbf{D}$ is open, as we have assumed, all global optima are also local optima. If $\mathbf{D}$ is closed, however, this might not be the case. In that case, a global optimum occurring at a point $\mathbf{x}^*$ on the boundary $\partial \mathbf{D}$ is not, strictly speaking, a local optimum, since a neighbourhood of $\mathbf{x}^*$ cannot be defined (see Fig. 2.1). Note that there also might be more than one global optimum. For example, the function $f(x) = \cos x$ has global minima at $x = \pi, 3\pi, \ldots$ and global maxima at $x = 0, 2\pi, \ldots$.

2.1.2 Objective functions

In optimization, the function f is referred to as the **objective function**[2] and the variables $\mathbf{x}$ are sometimes referred to as **decision variables**. The goal of optimization is to find either the global minima or the global maxima of the objective function. Whether one should search for the minima (minimization) or the maxima (maximization) depends on the problem in question, but it implies no restriction to consider only minimization, since the maximization of f is equivalent to the

---

[2] In connection with some optimization methods, the objective function might go under a different name. Thus, for example, in evolutionary algorithms, the objective function is usually referred to as the fitness function.

minimization of $-f$. In this chapter, we shall mainly consider minimization even though, in later chapters, maximization will be considered as well.

2.1.3 Constraints

In many optimization problems, the decision variables $\mathbf{x}$ are not allowed to take any value in $\mathbf{R}^n$ but are instead constrained by m **inequality constraints** given by

$$g_i(\mathbf{x}) \leq 0, \quad i = 1, \ldots, m, \tag{2.2}$$

and k **equality constraints** defined as

$$h_i(\mathbf{x}) = 0, \quad i = 1, \ldots, k. \tag{2.3}$$

g_i and h_i are referred to as **constraint functions**. Those points $\mathbf{x} \in \mathbf{R}^n$ that satisfy all the $m+k$ constraints are referred to as **feasible points**. The corresponding optimization problem, denoted as

$$\text{minimize} f(\mathbf{x}), \quad \mathbf{x} \in \mathbf{S}, \tag{2.4}$$

where $\mathbf{S} \subseteq \mathbf{R}^n$, the set of feasible points, is defined by the $m+k$ constraints, is referred to as a **constrained optimization problem**. In case $m = k = 0$, the problem is **unconstrained**. Note that the definition of the inequality constraints in eqn (2.2) implies no restriction, since a constraint of the form $g(\mathbf{x}) \geq 0$ can be rewritten as $-g(\mathbf{x}) \leq 0$. In fact, even the equality constraints can be written as a combination of two inequality constraints by noting that

$$h(\mathbf{x}) = 0 \Leftrightarrow \{h(\mathbf{x}) \leq 0 \text{ and } -h(\mathbf{x}) \leq 0\}. \tag{2.5}$$

2.2 Taxonomy of optimization problems

Optimization problems can be divided into categories in several different ways, depending on the properties of the objective function and the properties of the set $\mathbf{S}$ of feasible points. The distinction between unconstrained and constrained optimization has already been introduced above. One can also define **continuous optimization problems**, in which the objective function f and the constraint functions g_i and h_i are continuous, and **differentiable optimization problems**, in which the objective and constraint functions are continuously differentiable.

A common special case of continuous optimization is **linear programming**, in which the objective function is linear, and the constraint functions are affine, i.e. of the form $g(\mathbf{x}) = \mathbf{a}^T \mathbf{x} + b$, where $\mathbf{a}$ and $\mathbf{x}$ are n-dimensional vectors. More formally, linear programming is defined as

$$\left. \begin{array}{ll} \text{minimize} & \mathbf{c}^T \mathbf{x} \\ \text{subject to} & \mathbf{a}_i^T \mathbf{x} + b_i \leq 0, \quad i = 1, \ldots, m \\ \text{and} & -x_i \leq 0, \quad i = 1, \ldots, n. \end{array} \right\} \tag{2.6}$$

The restriction to non-negative variables is motivated by the fact that integer programming usually concerns problem involving, for example, production, transportation or investment, in which the quantities involved are, of course, non-negative. If the objective function or at least one of the constraint functions are non-linear, the problem is one of **non-linear programming**. An important special case of non-linear programming is **quadratic programming**, which, in vector notation, takes the form

$$\left. \begin{array}{ll} \text{minimize} & \frac{1}{2}\mathbf{x}^T\mathbf{Q}\mathbf{x} + \mathbf{c}^T\mathbf{x}, \\ \text{subject to} & \mathbf{a}_i^T\mathbf{x} + b_i \leq 0, \quad i = 1, \ldots, m \\ \text{and} & -x_i \leq 0, \quad i = 1, \ldots, n. \end{array} \right\} \quad (2.7)$$

where $\mathbf{Q}$ is a symmetric $n \times n$ matrix.

Another important class of optimization problems are those in which (at least one of the) the variables x_i take values in $\mathbf{Z}$. In the special case of **integer programming** [81], $\mathbf{x} \in \mathbf{Z}^n$, and in **binary integer programming**, $\mathbf{x} \in \{0, 1\}^n$. In **combinatorial optimization**, the goal, loosely speaking, is to find the minimum of a function f defined over a discrete set X of feasible points. A typical example is the travelling salesman problem (which will be studied thoroughly in later chapters), where the objective is to find the shortest possible path between n cities, such that each city is visited once, and only once, except for the city of origin that is revisited in the final step of the tour. Here, the set X can be defined as all possible permutations (i.e. orderings) of the cities c_i (or, rather, their indices, i). In addition to the travelling salesman problem, there are many other important problems in which the variables take discrete values, such as scheduling problems, assignment problems (assigning N people to carry out M jobs), and some investment problems (where decision variables take the value 1 if an investment is made, and 0 if it is not). In later chapters, we will encounter integer programming problems. However, for the remainder of this chapter, we will focus on continuous optimization problems.

2.3 Continuous optimization

As mentioned above, continuous optimization problems involve continuous functions, and differentiable optimization problems are thus a special case. In our study of continuous optimization, and particularly in the examples and exercises, we shall mostly be concerned with this special case, since many classical optimization methods, as we shall see below, rely on the computation of derivatives.

2.3.1 Properties of local optima

Local optima of differentiable functions can be found by studying the properties of the derivative. Consider the function $f: \mathbf{D} \to \mathbf{R}$, where $\mathbf{D} \subseteq \mathbf{R}$ is an open set. **A critical point** x^* of the function occurs where either (1) the derivative $f'(x^*) \equiv f'(x)|_{x=x^*}$ is equal to zero, in which case x^* is referred to as **a stationary point** or (2) the derivative does not exist. As is well known from basic calculus, the local optima

CLASSICAL OPTIMIZATION 13

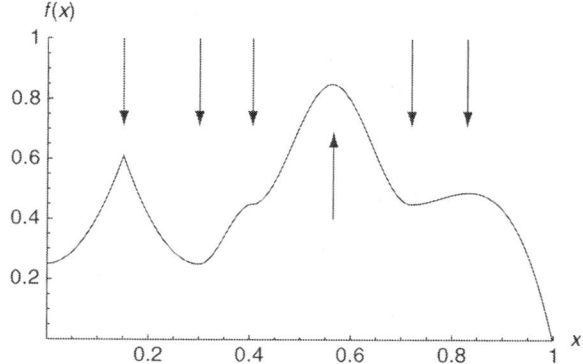

Figure 2.2: *An illustration of the critical points for a function $f(x)$ defined on the open interval $x \in]0,1[$. The leftmost critical point occurs at a cusp, where the derivative is not defined. The critical point near $x = 0.4$ is a saddle point, i.e. neither a local maximum nor a minimum.*

(minima and maxima) of f occur at critical points. The nature of stationary points can be studied by considering the sign of the first derivative around the stationary point. If there exist a $\delta > 0$ such that (1) $f(x)$ is continuous in $I(x^*, \delta)$, (2) the derivative $f'(x)$ exists in $I(x^*, \delta)$, (3) $f'(x) < 0 \ \forall x \in]x^* - \delta, x^*[$ and (4) $f'(x) > 0 \ \forall x \in]x^*, x^* + \delta[$, then f has a local minimum at x^*. Local maxima can be found in an analogous way, namely if $f'(x) > 0 \forall x \in]x^* - \delta, x^*[$ and $f'(x) < 0 \ \forall x \in]x^*, x^* + \delta[$. If neither condition holds, x^* is neither a local minimum nor a local maximum. An example is the function $f(x) = x^3$ which has a **saddle point** at $x = 0$. The function f in Fig. 2.2 was defined on the open set $\mathbf{D} = \{x: 0 < x < 1\}$ and even though f approaches zero near $x = 1$, it does not actually reach this value. Thus, all that can be said is that the **infimum** of f (denoted inf f) is equal to zero. However, had the boundary point been included in the domain $\mathbf{D}$ of f, the function would have attained its global minimum there. Thus, in optimization problems over closed sets, one must also investigate the function values on the boundary $\partial \mathbf{D}$ in order to make sure that one captures the global minimum (or maximum).

If the function f happens to be twice continuously differentiable in a neighbourhood $I(x^*, \delta)$ of x^*, the second derivative of f provides information about the stationary point: if there exist some $\delta > 0$ such that $f''(x) > 0 \ \forall x \in I(x^*, \delta)$, then f has a minimum at x^*. Similarly, if instead $f''(x) < 0 \ \forall x \in I(x^*, \delta)$, then f has a maximum at x^*.

The study of stationary points can, of course, be generalized to the multidimensional case as well. Let $f: \mathbf{R}^n \to \mathbf{R}$ and let

$$\nabla f = \left(\frac{\partial f}{\partial x_1}, \ldots, \frac{\partial f}{\partial x_n} \right)^\mathrm{T} \tag{2.8}$$

denote the **gradient** of f. The stationary points $\mathbf{x}^*$ of f occur where $\nabla f(\mathbf{x}^*) \equiv \nabla f(\mathbf{x})|_{\mathbf{x}=\mathbf{x}^*} = 0$. The nature of the stationary points of a twice continuously differentiable function $f: \mathbf{R}^n \to \mathbf{R}$ can be studied by considering the properties of the **Hessian**, defined as

$$H = \begin{pmatrix} \frac{\partial^2 f}{\partial x_1^2} & \frac{\partial^2 f}{\partial x_1 \partial x_2} & \cdots & \frac{\partial^2 f}{\partial x_1 \partial x_n} \\ \frac{\partial^2 f}{\partial x_2 \partial x_1} & \ddots & & \vdots \\ \vdots & & \ddots & \vdots \\ \frac{\partial^2 f}{\partial x_n \partial x_1} & \cdots & \cdots & \frac{\partial^2 f}{\partial x_n^2} \end{pmatrix} \tag{2.9}$$

Note that H is symmetric, since

$$\frac{\partial^2 f}{\partial x_i \partial x_j} = \frac{\partial^2 f}{\partial x_j \partial x_i} \quad \forall i, j \in 1, \ldots, n. \tag{2.10}$$

If H is **positive definite** at a stationary point $\mathbf{x}^*$, i.e. if all its eigenvalues are positive,[3] then $\mathbf{x}^*$ is a local minimum of f. Similarly, if H is **negative definite** at $\mathbf{x}^*$, then f has a local maximum. If some eigenvalues are negative and some are positive, $\mathbf{x}^*$ is a saddle point, i.e. neither a local maximum nor a local minimum. In the special case where $n = 2$, the eigenvalues $\lambda_{1,2}$ are obtained by solving the determinant equation

$$\begin{vmatrix} \frac{\partial^2 f}{\partial x_1^2} - \lambda & \frac{\partial^2 f}{\partial x_1 \partial x_2} \\ \frac{\partial^2 f}{\partial x_1 \partial x_2} & \frac{\partial^2 f}{\partial x_2^2} - \lambda \end{vmatrix} \equiv \left(\frac{\partial^2 f}{\partial x_1^2} - \lambda \right) \left(\frac{\partial^2 f}{\partial x_2^2} - \lambda \right) - \left(\frac{\partial^2 f}{\partial x_1 \partial x_2} \right)^2 = 0. \tag{2.11}$$

The study of critical points described above can be used for finding *local* optima of a function f. Global optima are another matter and, in general, local tests of the kind just discussed give no information regarding such optima. In other words, if a local minimum of a function f, defined on the open set $\mathbf{D} \subseteq \mathbf{R}^n$, has been found, one cannot say whether this point is also a *global* minimum. However, the special case of **convex functions**, which will be considered next, is an important exception.

2.3.2 Global optima of convex functions

2.3.2.1 Convex sets and functions
Consider a set $\mathbf{S} \subseteq \mathbf{R}^n$. $\mathbf{S}$ is **convex** if, for any pair of points $\mathbf{x}_1, \mathbf{x}_2 \in \mathbf{S}$, all points on the line connecting $\mathbf{x}_1$ to $\mathbf{x}_2$ also belong to $\mathbf{S}$. Formally, if

$$\mathbf{x}_1, \mathbf{x}_2 \in \mathbf{S} \Rightarrow a\mathbf{x}_1 + (1-a)\mathbf{x}_2 \in \mathbf{S}, \quad \forall a \in [0, 1] \tag{2.12}$$

---

[3] If all eigenvalues are non-negative (but not necessarily positive), then H is said to be **positive semi-definite**.

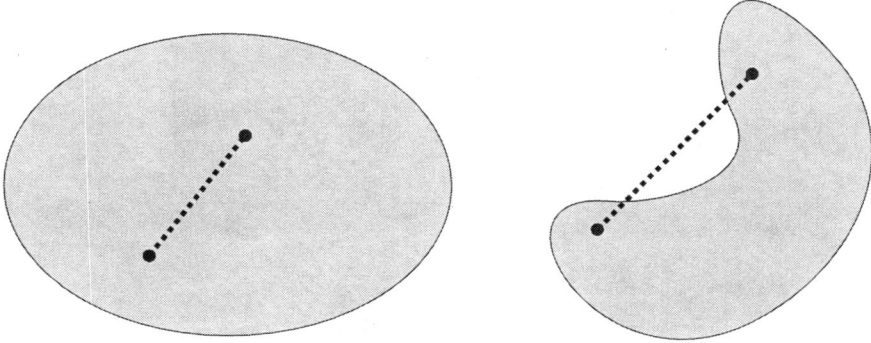

Figure 2.3: *Left panel: a convex set* $S \subseteq R^2$. *Right panel: a non-convex set.*

then S is convex. Fig. 2.3 shows an example of a convex set (left panel) and a non-convex set (right panel). Convexity can also be defined for functions: let $S \subseteq R^n$ be a convex set, and consider the function $f: S \to R$. f is said to be convex if, for any $x_1, x_2 \in S$, and for any $a \in [0, 1]$

$$f(ax_1 + (1-a)x_2) \leq af(x_1) + (1-a)f(x_2). \tag{2.13}$$

In addition, if

$$f(ax_1 + (1-a)x_2) < af(x_1) + (1-a)f(x_2), \tag{2.14}$$

for all $x_1 \neq x_2$ and for all $a \in\,]0, 1[$, then f is **strictly convex**. A simple example of a function that is strictly convex on $S = R$, namely $f(x) = x^2$ is given in Fig. 2.4. This figure also illustrates the geometrical interpretation of convexity. Loosely speaking, between any two points x_1 and x_2 in the domain of a convex function f, the line connecting the two points lies above the function.

Note that, if a function f is convex on a set S, it can be proven that the function is also continuous on (the interior of) S. If a function f, defined on an (open) convex set $S \subseteq R^n$ happens to be twice continuously differentiable, the Hessian H can be used to determine whether or not f is convex: it can be shown that if (and only if) H is positive semi-definite for all $x \in S$, then f is convex. Furthermore, if H is positive definite for all $x \in S$, then f is strictly convex. Note that, in the case of strict convexity, the implication is one-sided: not all strictly convex functions have a positive definite Hessian. For example, consider the function $f(x) = x^4$, on an open set S containing the point $x = 0$. $f(x)$ is strictly convex, yet the Hessian, which in the one-dimensional case reduces to $f''(x)$, is equal to zero at $x = 0$.

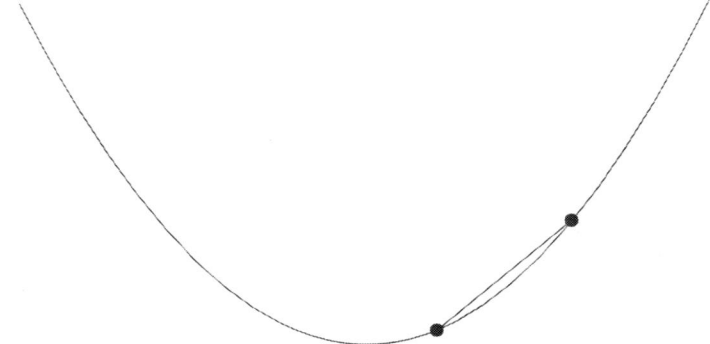

Figure 2.4: *The convex function $f(x) = x^2$.*

Example 2.1
Determine whether or not the function

$$f(x_1, x_2) = 2x_1^2 - 2x_1 x_2 + 2x_2^2 + 6x_1 - 6x_2 \qquad (2.15)$$

is convex over $\mathbf{R}^2$.

Solution Since the function is everywhere twice continuously differentiable, we can investigate the Hessian, which in this case takes the form

$$H = \begin{pmatrix} \frac{\partial^2 f}{\partial x_1^2} & \frac{\partial^2 f}{\partial x_1 \, \partial x_2} \\ \frac{\partial^2 f}{\partial x_1 \, \partial x_2} & \frac{\partial^2 f}{\partial x_2^2} \end{pmatrix} = \begin{pmatrix} 4 & -2 \\ -2 & 4 \end{pmatrix}, \qquad (2.16)$$

with eigenvalues $\lambda_{1,2} = 4 \pm 2$. Thus, the Hessian is positive definite, and the function is not only convex but strictly convex. ∎

2.3.2.2 Optima of convex functions

Returning to the left panel of Fig. 2.4, we note that the function $f(x) = x^2$ has a single local minimum at $x^* = 0$, and that this point also is the *global* minimum of the function. In fact, this is a general property of convex functions: one can prove that, if f is convex, any local optimum will also be a global optimum. A proof of this property is given in Appendix B. Furthermore, if f is strictly convex (as, for example, in the case of $f(x) = x^2$), then the global optimum is unique. Evidently, convexity is a very desirable property in optimization, and one that is frequently exploited in classical optimization [5]. In particular, there exist efficient numerical methods for solving so-called convex optimization problems, some of which will be considered briefly in the next section.

2.4 Algorithms for continuous optimization

Having introduced some properties of continuous optimization problems, we will now turn to the issue of algorithms for solving such problems. The taxonomy of such algorithms can be constructed in different ways. For example, one may distinguish between analytical and numerical (typically iterative) methods. However, here, we shall make the division with respect to problem type, distinguishing between unconstrained and constrained optimization, starting with the former. Without loss of generality, we shall only consider *minimization*, and it will further be assumed that the functions under consideration are at least twice continuously differentiable over $\mathbf{R}^n$.

2.4.1 Unconstrained optimization

In unconstrained optimization, $m = k = 0$ in eqns (2.2) and (2.3), so that $\mathbf{S} = \mathbf{R}^n$ and the problem can therefore be written as

$$\text{minimize} f(\mathbf{x}), \quad \mathbf{x} \in \mathbf{R}^n. \tag{2.17}$$

A number of iterative algorithms have been developed for the solution of such problems. In these methods, one generates a sequence of iterates $\mathbf{x}_j$ that gradually converge towards a local optimum, beginning at a starting point $\mathbf{x}_0$. In essence, the algorithms generate new iterates as

$$\mathbf{x}_{j+1} = \mathbf{x}_j + \eta_j \mathbf{d}_j, \tag{2.18}$$

where $\mathbf{d}_j$ is the search direction at the point $\mathbf{x}_j$, and η_j is the step length. Algorithms differ regarding the manner in which the search direction is chosen. The step length is typically adaptive, i.e. its value also depends on $\mathbf{x}$. The determination of an appropriate step length usually requires a one-dimensional **line search**, and this will be our first topic.

2.4.1.1 Line search
As is clear from eqn (2.18), once a search direction $\mathbf{d}_j$ has been specified, the new iterate $\mathbf{x}_j$ will depend only on the parameter η (for simplicity we will drop the subscript j on η_j). Inserting the new iterate $\mathbf{x}_{j+1}$ in the objective function $f(\mathbf{x})$, one obtains the optimization problem

$$\text{minimize} f(\mathbf{x}_{j+1}(\eta)) \Leftrightarrow \text{minimize } \phi(\eta), \tag{2.19}$$

with $\eta \in \mathbf{R}$. Thus, for each new search direction $\mathbf{d}_j$ one obtains a one-dimensional optimization problem, also referred to as a line-search problem, which is really only a special case of the general unconstrained optimization problem. In simple cases, the line-search problem can be solved analytically. If not, the problem can be attacked using an iterative method. Here, a single example will be given, namely the **bisection method**. However, one-dimensional cases of the general optimization methods (e.g. Newton's method) presented below can, of course, also be used for line search.

1. Start with a sufficiently large interval $[a_0, b_0]$, such that it includes the (local) minimum η^*.
2. Compute the derivative $\phi'(c_j)$ (where $j = 0$ in the first iteration) at the centre point $c_j = (a_j + b_j)/2$.
3. If $\phi'(c_j) < 0$, set $a_{j+1} = c_j$ and $b_{j+1} = b_j$. If instead $\phi'(c_j) > 0$, set $a_{j+1} = a_j$ and $b_{j+1} = c_j$.
4. Repeat steps 2 and 3 until convergence.
5. Finally, set $\eta^* = c_k$, where k denotes the index of the final iteration.

Algorithm 2.1: *The bisection method.*

2.4.1.1.1 The bisection method This method, also known as **interval halving**, finds a (local) minimum by successively reducing the size of the initial search interval $[a_0, b_0]$ based on the properties of the derivative $\phi'(\eta)$. Finding an appropriate initial interval is not always trivial. However, as the aim is to find a point η^* such that $\phi'(\eta^*) = 0$, it makes sense to start with points a_0 and b_0 such that $\phi'(a_0) < 0$ and $\phi'(b_0) > 0$. The method is presented as Algorithm 2.1. Since the size of the search interval is reduced by half in each iteration, the number of iterations k needed in order to find η^* with an error not exceeding γ is given by

$$\left(\frac{1}{2}\right)^k = \frac{\gamma}{b_0 - a_0}. \tag{2.20}$$

Example 2.2
Consider the problem of minimizing the function

$$f(\eta) = e^\eta - 2\eta. \tag{2.21}$$

using the bisection method, with the starting interval $[a_0, b_0] = [0, 2]$.

Solution As is evident from inspection of the function, there is a minimum in this range. Since $f''(\eta) = e^\eta > 0$, the function is strictly convex (remember that the Hessian is reduced to the second derivative for a one-dimensional function!), so that this is also the global minimum of $f(\eta)$. The derivative $f'(\eta)$ equals $e^\eta - 2$, and it is thus elementary to determine the exact location of the minimum as

$$\eta^* = \ln 2 \approx 0.693. \tag{2.22}$$

Nevertheless, let us apply line search. Evidently, $f'(a_0) < 0$ and $f'(b_0) > 0$. In the first iteration of the line search, the centre of the interval is obtained as

$$c_0 = \frac{a_0 + b_0}{2} = 1. \tag{2.23}$$

Hence, $f'(c_0) = f'(1) = e^1 - 2 > 0$. Thus, we set $a_1 = a_0 = 0$ and $b_1 = c_0 = 1$. In the second iteration, $c_1 = 1/2$, and $f'(c_1) = e^{\frac{1}{2}} - 2 < 0$, so that $a_2 = c_1 = 1/2$ and $b_2 = b_1 = 1$, etc. The points $\{c_0, c_1, \ldots, c_6\}$ are shown in Fig. 2.5. $c_6 \approx 0.695$ and $f'(c_6) \approx 0.004$. ∎

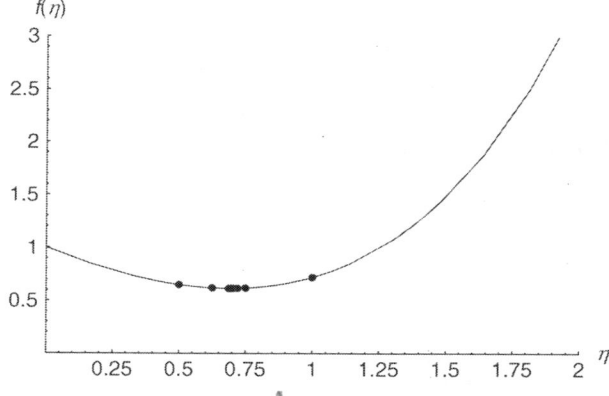

Figure 2.5: *An illustration of a line search (see Example 2.2) that finds the minimum of the function $f(\eta) = e^\eta - 2\eta$. The points $(c_k, f(c_k))^T$ obtained in the iterative procedure are superposed on the graph.*

We shall now proceed to algorithms for the general, n-dimensional optimization problem, eqn (2.17).

2.4.1.2 Gradient descent

Making a Taylor expansion of $f(\mathbf{x})$ at $\mathbf{x} = \mathbf{x}_0$, and neglecting higher-order terms, one obtains

$$f(\mathbf{x}) \approx f(\mathbf{x}_0) + \nabla f(\mathbf{x}_0)^T(\mathbf{x} - \mathbf{x}_0). \qquad (2.24)$$

Since the aim is to minimize $f(\mathbf{x})$, starting from $\mathbf{x} = \mathbf{x}_0$, we should try to find an $\mathbf{x}$ such that $f(\mathbf{x}) < f(\mathbf{x}_0)$. There might be many search directions for which this condition holds, but the direction in which $f(\mathbf{x})$ decreases *most* is given by the negative gradient at $\mathbf{x}_0$ (the simple proof is given in Appendix B, Section B.1.2). Thus, given only gradient information as in eqn (2.24), a suitable choice of search direction is given by

$$\mathbf{d}_0 = -\nabla f(\mathbf{x}_0). \qquad (2.25)$$

Returning to eqn (2.18) the iteration now takes the form

$$\mathbf{x}_{j+1} = \mathbf{x}_j - \eta_j \nabla f(\mathbf{x}_j). \qquad (2.26)$$

This is the method of **gradient descent**. The step length η_j is typically determined using line search, as described above. The method is summarized in Algorithm 2.2.

In cases where the objective function has a very complex shape, the line-search problem may itself be computationally very expensive, and one may then resort to using a fixed step length η (typically determined through empirical tests) or a slowly decreasing η. A procedure for modification of the step length is considered in connection with the backpropagation algorithm (which is a form of gradient descent)

1. Select a starting point $\mathbf{x}_0$.
2. Compute the gradient $\nabla f(\mathbf{x}_j)$ (where $j = 0$ in the first iteration) and form the quantity $\mathbf{x}_j - \eta_j \nabla f(\mathbf{x}_j) \equiv \phi(\eta_j)$, which is a function only of the variable η_j.
3. Using line search, e.g. by means of the bisection method defined in Algorithm 2.1, find the step length $\eta_j = \eta^*$ that minimizes $\phi(\eta_j)$, and set $\mathbf{x}_{j+1} = \mathbf{x}_j - \eta^* \nabla f(\mathbf{x}_j)$.
4. Repeat steps 2 and 3 until convergence, i.e. until the gradient vanishes.

Algorithm 2.2: *The gradient descent algorithm.*

in Appendix A, see eqn (A27). The gradient descent algorithm typically generates a zig-zag search path towards the local minimum. In fact, one can show that, under gradient descent, consecutive search directions are orthogonal to each other and therefore other algorithms that take a more direct path towards the minimum have been formulated. One such algorithm is the **conjugate gradient** method, which forms the new search direction as a combination of the current gradient and the gradient obtained in the previous iteration. Such methods are beyond the scope of this text, however. Let us instead turn to an example of gradient descent.

Example 2.3
Gradient descent is to be applied to the problem of minimizing the function

$$f(x_1, x_2) = x_1^2 - x_1 x_2 + x_2^2, \qquad (2.27)$$

shown in Fig. 2.6, starting from the point $\mathbf{x}_0 = (-1, 1)^T$. Determine the descent direction at $\mathbf{x}_0$, as well as the appropriate step length η. Find also the point $\mathbf{x}_1$ resulting from the first iteration.

Solution The gradient of f is easily obtained as

$$\nabla f(\mathbf{x}) = \left(\frac{\partial f}{\partial x_1}, \frac{\partial f}{\partial x_2} \right)^T = (2x_1 - x_2, -x_1 + 2x_2)^T. \qquad (2.28)$$

Thus, at $(-1, 1)^T$ the search direction $\mathbf{d}_0$ becomes

$$\mathbf{d}_0 = -\nabla f(\mathbf{x}_0) = (3, -3)^T. \qquad (2.29)$$

The new point $\mathbf{x}_1$ thus becomes

$$\mathbf{x}_1 = \mathbf{x}_0 - \eta \nabla f(\mathbf{x}_0) = (-1, 1)^T + \eta(3, -3)^T = (-1 + 3\eta, 1 - 3\eta)^T, \qquad (2.30)$$

where, for convenience of notation, the subscript has been dropped from η_0. Inserting $\mathbf{x}_1$ in the objective function one obtains

$$\phi(\eta) \equiv f(-1 + 3\eta, 1 - 3\eta) = \cdots = 3(1 - 3\eta)^2. \qquad (2.31)$$

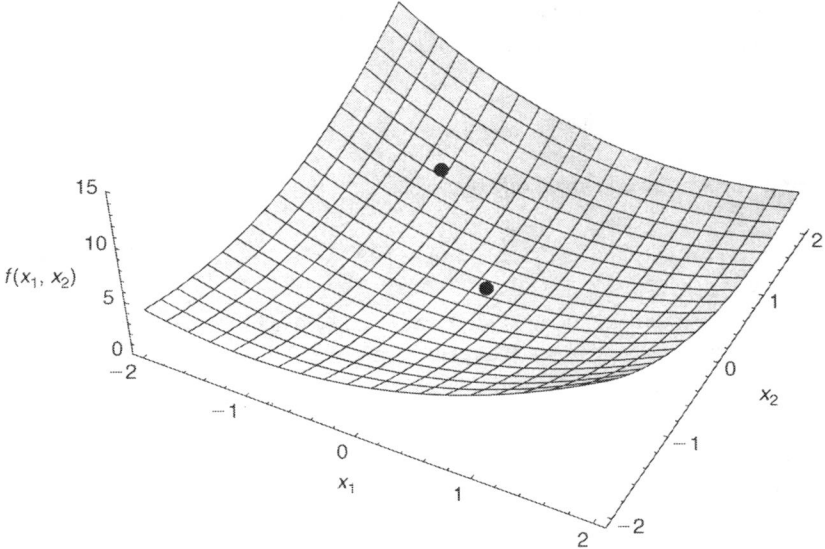

Figure 2.6: *The function $f(x_1, x_2) = x_1^2 - x_1 x_2 + x_2^2$ (see Example 2.3). The starting point $\mathbf{x}_0 = (-1, 1)^T$ and the point $\mathbf{x}_1$, reached after one iteration of gradient descent, are both indicated in the figure.*

The step length should be chosen so as to minimize $\phi(\eta)$. In this simple case, not even a line search is needed: it is evident from eqn (2.31) that the minimum of $\phi(\eta)$ occurs at $\eta^* = 1/3$, so that $\mathbf{x}_1 = (0, 0)^T$. ∎

2.4.1.3 Newton's method

By contrast with gradient descent, **Newton's method** uses information concerning the second derivative of $f(\mathbf{x})$. Retaining second-order terms in the Taylor expansion of $f(\mathbf{x})$, one obtains

$$f(\mathbf{x}) \approx f(\mathbf{x}_0) + \nabla f(\mathbf{x}_0)^T (\mathbf{x} - \mathbf{x}_0) + \frac{1}{2}(\mathbf{x} - \mathbf{x}_0)^T H(\mathbf{x}_0)(\mathbf{x} - \mathbf{x}_0), \quad (2.32)$$

where $H(\mathbf{x}_0)$ is the Hessian matrix, defined in eqn (2.9). Starting with the one-dimensional case, eqn (2.32) is reduced to

$$f(x) \approx f(x_0) + f'(x_0)(x - x_0) + \frac{1}{2} f''(x_0)(x - x_0)^2 \equiv f_{[2]}(x), \quad (2.33)$$

where $f_{[2]}(x)$ denotes the second-order Taylor expansion of $f(x)$. As in the gradient descent method described above, the aim is to take a step from x_0 to x so as to obtain the lowest-possible value of $f(x)$. In other words, one should attempt to find the minimum of $f_{[2]}(x)$, and use the corresponding point x^* as the starting point x_1

for the next iteration. Now, minima are found at stationary points that, in turn, are found where the derivative vanishes. Thus, the minimum is obtained as the solution to the linear equation

$$f'_{[2]}(x) = 0 \Leftrightarrow f'(x_0) + (x - x_0)f''(x_0) = 0. \tag{2.34}$$

Solving this simple equation, one obtains

$$x^* = x_0 - \frac{f'(x_0)}{f''(x_0)}. \tag{2.35}$$

The one-dimensional version just described can be used for line search, and it also goes under the name **Newton–Rhapson's method**. The method is summarized in Algorithm 2.3.

---

1. Select a starting point x_0.
2. Form the iterate x_{j+1} as

$$x_{j+1} = x_j - \frac{f'(x_j)}{f''(x_j)}$$

where $j = 0$ in the first iteration.
3. Repeat step 2 until convergence.

Algorithm 2.3: *Newton–Rhapson's method.*

Returning to the general case, the second-order Taylor expression can be written

$$f_{[2]}(\mathbf{x}) = f(\mathbf{x}_0) + \nabla f(\mathbf{x}_0)^T (\mathbf{x} - \mathbf{x}_0) + \frac{1}{2}(\mathbf{x} - \mathbf{x}_0)^T H(\mathbf{x}_0)(\mathbf{x} - \mathbf{x}_0), \tag{2.36}$$

and the gradient of $f_{[2]}(\mathbf{x})$ is given by

$$\nabla f_{[2]}(\mathbf{x}) = \nabla f(\mathbf{x}_0) + H(\mathbf{x}_0)(\mathbf{x} - \mathbf{x}_0), \tag{2.37}$$

so that, in analogy with the one-dimensional case above, the minimum is found at

$$\mathbf{x}^* = \mathbf{x}_0 - H^{-1}(\mathbf{x}_0)\nabla f(\mathbf{x}_0). \tag{2.38}$$

Thus, in the general case, Newton's method forms new iterates according to

$$\mathbf{x}_{j+1} = \mathbf{x}_j - H^{-1}(\mathbf{x}_j)\nabla f(\mathbf{x}_j). \tag{2.39}$$

Note that the step length η_j is absent in eqn (2.39). In **Newton's modified method** the step length is included, so that

$$\mathbf{x}_{j+1} = \mathbf{x}_j - \eta_j H^{-1}(\mathbf{x}_j)\nabla f(\mathbf{x}_j), \tag{2.40}$$

where η_j again can be determined using line search.

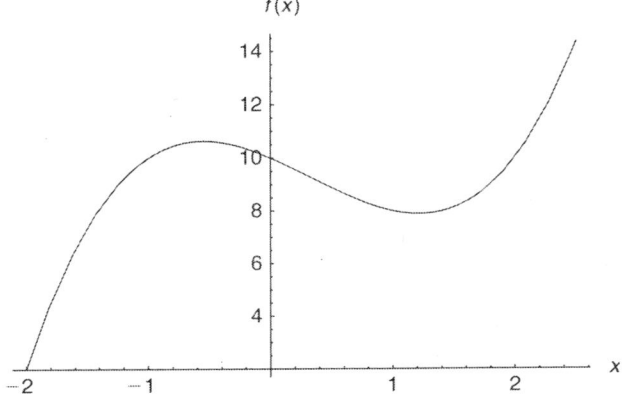

Figure 2.7: *The function $f(x) = x^3 - x^2 - 2x + 10$, considered in Example 2.4.*

We should pause here to note that, even though we have used the terms *minima* and *minimization* above, there is no guarantee that Newton's method always converges towards a minimum: the method is based on finding stationary points, of any kind, of the second-order Taylor expansion of $f(\mathbf{x})$, and if $\mathbf{x}_0$ is chosen badly, the direction provided by eqn (2.39) may lead to a maximum or a saddle point (see Example 2.4). In fact, in order for Newton's method to converge towards a minimum, $f(\mathbf{x})$ must be convex in a neighbourhood of $\mathbf{x}_0$.[4]

Example 2.4
Consider the function

$$f(x) = x^3 - x^2 - 2x + 10, \tag{2.41}$$

shown in Fig. 2.7. As is obvious from the figure, the function has two stationary points in the range $]-2, 2[$, and they are located at

$$x^*_{1,2} = \frac{1}{3}(1 \pm \sqrt{7}). \tag{2.42}$$

Now, using Newton–Rhapson's method, the iteration becomes

$$x_{j+1} = x_j - \frac{f'(x_j)}{f''(x_j)} = x_j - \frac{3x_j^2 - 2x_j - 2}{6x_j - 2}. \tag{2.43}$$

Thus, starting from, say, $x_0 = 2$, one obtains the sequence $x_1 = 1.400$, $x_2 = 1.231$, $x_3 = 1.215, \ldots$, which converges towards the local minimum. However, if instead the starting point is chosen as $x_0 = 0$, the sequence $x_1 = -1.000$, $x_2 = -0.625$, $x_3 = -0.5516$,

---

[4] In the special case of a quadratic, convex function, one can show that the method will converge in a single step.

$x_4 = -0.5486, \ldots$ is obtained. Thus, in the second case, the sequence converges towards the local maximum. ∎

In addition to the convergence problems just mentioned, finding the inverse of the Hessian is often a computationally very expensive procedure. Thus, a variety of modified methods that rely on approximate computation of the inverse of the Hessian have been suggested, such as **quasi–Newton methods** and the **Lewenberg–Marquardt algorithm**, both of which are beyond the scope of this text.

2.4.2 Constrained optimization

Most of the optimization problems occurring in realistic applications are associated with some form of constraints. In such cases, the problem takes the general form

$$\left. \begin{array}{ll} \text{minimize} & f(\mathbf{x}) \\ \text{subject to} & g_i(\mathbf{x}) \leq 0, \quad i = 1, \ldots, m \\ \text{and} & h_i(\mathbf{x}) = 0, \quad i = 1, \ldots, k. \end{array} \right\} \quad (2.44)$$

Again, we shall assume that both the objective function and the constraint functions are at least twice continuously differentiable on $\mathbf{R}^n$. In simple cases, a constrained optimization problem can be converted to an unconstrained one, as shown in Example 2.5.

Example 2.5
Consider the problem

$$\left. \begin{array}{ll} \text{minimize} & f(x_1, x_2) = x_1^4 - x_2^2 \\ \text{subject to} & x_1 + x_2 = 2 \end{array} \right\} \quad (2.45)$$

Reduce the problem to an unconstrained problem of a single variable.

Solution In this simple case, one of the variables (e.g. x_2) can be removed by rewriting the equality constraint as $x_2 = 2 - x_1$, and inserting the result in the objective function, so that the problem takes the form

$$\text{minimize} \quad \tilde{f}(x_1) = x_1^4 - (2 - x_1)^2 \quad (2.46)$$

with $x_1 \in \mathbf{R}$. ∎

Commonly, however, the constraints cannot be removed so easily, and one is thus faced with the problem of somehow coping with them. There is no single method for solving the general problem, in which the objective function and the constraints may take any form, even with the restriction to twice continuously differentiable functions. However, there are many important special cases for which efficient methods are available. In particular, for **convex optimization problems**, in which both the objective function $f(\mathbf{x})$ and all inequality constraints $g_i(\mathbf{x})$ are convex functions, and the equality constraints are **affine**, i.e. take the form

$$h_i(\mathbf{x}) = \mathbf{A}_i^T \mathbf{x} + b_i \quad i = 1, \ldots, k, \quad (2.47)$$

any local optimum will also be a global optimum. This is so, since the set $\mathbf{S} \subseteq \mathbf{R}^n$ defined by convex inequality constraints and affine equality constraints is convex, so that the proof in Appendix B, Section B.1.1, still holds. Note that the equality constraints must not only be convex, but also affine in order for to $\mathbf{S}$ to be necessarily convex. To make this probable, consider the minimization of a function $f(x_1, x_2, x_3)$ subject to the constraints $x_1^2 + x_2^2 + x_3^2 \leq 2$ and $x_1 = 1$. Clearly, the equality constraint can easily be removed, and the inequality constraint then takes the form $x_2^2 + x_3^2 \leq 1$, which defines a convex set. On the other hand, if the equality constraint is, say, given by $x_1^2 = 1$, its elimination results in the set $\mathbf{S} = \{x_2^2 + x_3^2 \leq 1, x_1 = \pm 1\}$, which most certainly is not convex!

Several methods have been developed for the numerical solution of convex optimization problems. Examples include the **Simplex method** applicable to linear programming problems (see eqn (2.6)), the **Frank–Wolfe algorithm**, applicable to quadratic programming problems (see eqn (2.7)), **subgradient methods**, and **interior point methods**. Here, we shall contend ourselves with a brief discussion of interior point methods. First, however, analytical methods for constrained optimization will be considered.

2.4.2.1 The method of Lagrange multipliers

Consider a constrained optimization problem with only equality constraints, e.g. $m = 0, k > 0$ in eqn (2.44). Simplifying further, let us begin by studying the case $n = 2$, $k = 1$. In this case, the optimization problem takes the form

$$\left. \begin{array}{ll} \text{minimize} & f(x_1, x_2) \\ \text{subject to} & h(x_1, x_2) = 0 \end{array} \right\} \quad (2.48)$$

Since the problem is two-dimensional, the level curves (contour lines) of both f and h can be visualized as in Fig. 2.8. Now, consider a movement along one of the level curves $h(x_1, x_2) = 0$. As we move along this curve the value of $f(x_1, x_2)$ may, of course, change (increase or decrease). Indeed, at the point P in the figure, we can decrease the function value $f(x_1, x_2)$ by moving (inward) along the level curve $h(x_1, x_2) = 0$. On the other hand, at the point Q, where the level curve of f is parallel to that of h, any movement along $h(x_1, x_2) = 0$ will result in an increase in f. The point Q is thus a local minimum of f, subject to the constraint $h(x_1, x_2) = 0$. Furthermore, at any point $\mathbf{x} = (x_1, x_2)^\mathrm{T}$, ∇f is perpendicular to the level curve of f, and ∇h is likewise perpendicular to the level curve of h. Thus, at the minimum point Q, the two gradients must be parallel (since the level curves are parallel at that point). In other words,

$$\nabla f(\mathbf{x}_Q) = -\lambda \nabla h(\mathbf{x}_Q) \quad (2.49)$$

where $\mathbf{x}_Q$ denotes the point Q, and the minus sign is introduced for convenience of notation (see below). The informal argument that led to eqn (2.49) can be made rigorous, and the realization that the two gradients must be parallel at the optima

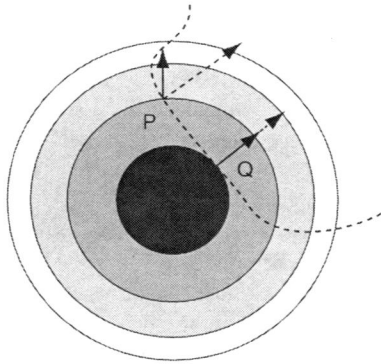

Figure 2.8: *The solid lines show the contours $f(x_1, x_2) = c$, for some function f and for different values of c, with the lowest values of c occurring near the centre of the figure. The dotted line shows the contour for the constraint $h(x_1, x_2) = 0$. The gradients of f (solid arrows) and h (dotted arrows) are shown at two different points. Note that the gradients of f and h are parallel at the local minimum of f, subject to the constraint.*

of f underlies the **Lagrange multiplier method**. Define $L(x_1, x_2, \lambda)$ as

$$L(x_1, x_2, \lambda) = f(x_1, x_2) + \lambda h(x_1, x_2). \tag{2.50}$$

Setting the gradient of L equal to zero, we obtain

$$\frac{\partial L}{\partial x_1} = \frac{\partial f}{\partial x_1} + \lambda \frac{\partial h}{\partial x_1} = 0, \tag{2.51}$$

$$\frac{\partial L}{\partial x_2} = \frac{\partial f}{\partial x_2} + \lambda \frac{\partial h}{\partial x_2} = 0, \tag{2.52}$$

$$\frac{\partial L}{\partial \lambda} = h = 0. \tag{2.53}$$

Thus, the first two equations summarize the requirement, given in eqn (2.49) that the gradients of f and h should be parallel at a local optimum $\mathbf{x}^*$ of the constrained optimization problem, whereas the final equation is a restatement of the constraint. Now, since critical points (of a continuously differentiable function) occur where the gradient of the function vanishes, the equations above indicate that the optima of the function $f(x_1, x_2)$, subject to the constraint $h(x_1, x_2)$ occur at the stationary points of L. The Lagrange multiplier method can thus be formulated as follows: if the function $f(x_1, x_2)$ has a local optimum at $(x_1^*, x_2^*)^T$ under the equality constraint $h(x_1, x_2) = 0$ and at least one of the derivatives $\partial h / \partial x_1$ and $\partial h / \partial x_2$ are $\neq 0$ at $(x_1^*, x_2^*)^T$, then there exist a λ, called the **Lagrange multiplier**, such that $(x_1^*, x_2^*, \lambda)^T$ is a stationary point

CLASSICAL OPTIMIZATION 27

of the function $L(x_1, x_2, \lambda)$ defined in eqn (2.50). Note, however, that there is no guarantee that the optimum $(x_1^*, x_2^*)^T$ is a minimum. Thus, the nature of the optima must be studied *a posteriori*.

The Lagrange multiplier method can be generalized to arbitrary dimensions and arbitrary number of equality constraints. In general, the optima of $f(\mathbf{x})$, subject to k equality constraints of the form $h_i(\mathbf{x}) = 0$, occur at the stationary points of $L(\mathbf{x}, \lambda)$, defined as

$$L(\mathbf{x}, \lambda) = f(\mathbf{x}) + \sum_{i=1}^{k} \lambda_i h_i(\mathbf{x}). \qquad (2.54)$$

The method is summarized in Algorithm 2.4.

---

1. Starting from the objective function $f(\mathbf{x})$ and the equality constraints $h_i(\mathbf{x})$, $i = 1, \ldots, k$, generate the function $L(\mathbf{x}, \lambda)$ as

$$L(\mathbf{x}, \lambda) = f(\mathbf{x}) + \sum_{i=1}^{k} \lambda_i h_i(\mathbf{x}).$$

2. Find the stationary points of L.
3. Investigate the stationary points, one by one, to find the minima (or maxima).

Algorithm 2.4: *The Lagrange multiplier method.*

Example 2.6
Use the Lagrange multiplier method to find the minimum of the function

$$f(x_1, x_2) = x_1 x_2^2 \qquad (2.55)$$

(see Fig. 2.9), subject to the constraint

$$x_1^2 + x_2^2 - 1 = 0. \qquad (2.56)$$

Solution Introducing the Lagrange multiplier λ, we can form the function L as

$$L(x_1, x_2, \lambda) = f(x_1, x_2) + \lambda h(x_1, x_2) = x_1 x_2^2 + \lambda \left(x_1^2 + x_2^2 - 1 \right). \qquad (2.57)$$

Thus, stationary points of L occur where the following three equations hold:

$$\frac{\partial L}{\partial x_1} = x_2^2 + 2\lambda x_1 = 0, \qquad (2.58)$$

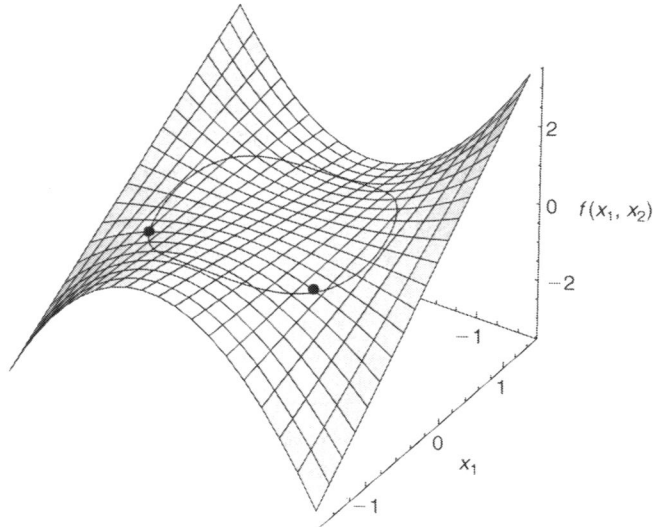

Figure 2.9: *The surface defined by the function $f(x_1,x_2) = x_1 x_2^2$, shown together with the constraint curve $h(x_1,x_2) = x_1^2 + x_2^2 - 1 = 0$. The dots indicate the location of the two minima of $f(x_1,x_2)$, subject to the constraint $h(x_1,x_2) = 0$.*

$$\frac{\partial L}{\partial x_2} = 2x_1 x_2 + 2\lambda x_2 = 0, \quad (2.59)$$

$$\frac{\partial L}{\partial \lambda} = x_1^2 + x_2^2 - 1 = 0. \quad (2.60)$$

From eqn (2.59) it follows that either $x_2 = 0$ or $\lambda = -x_1$. If $x_2 = 0$, the constraint equation gives $x_1 = \pm 1$ and, therefore, $\lambda = 0$. On the other hand, if $\lambda = -x_1$ then eqn (2.58) becomes

$$x_2^2 - 2x_1^2 = 0, \quad (2.61)$$

so that, in combination with the constraint equation, we obtain $x_1 = \pm 1/\sqrt{3}$, $x_2 = \pm\sqrt{2/3}$. Hence, there are six critical points, for which the function f takes the value $f(\pm 1, 0) = 0$, $f(1/\sqrt{3}, \pm\sqrt{2/3}) = 2/(3\sqrt{3})$, $f(-1/\sqrt{3}, \pm\sqrt{2/3}) = -2/(3\sqrt{3})$. Thus, the minima occur at the points $(-1/\sqrt{3}, \pm\sqrt{2/3})^T$. ∎

The discussion above concerned equality constraints. However, the method of Lagrange multipliers can be generalized to problems involving inequality constraints as well. If an optimization problem involves inequality constraints, the so called **Karush–Kuhn–Tucker (KKT) conditions** provide necessary conditions for a local optimum (in the general case) as well as necessary and sufficient conditions for a global optimum (in the case of convex optimization). While the formulation of the KKT conditions are beyond the scope of this text, we shall next consider an analytical method for minimization under inequality constraints.

2.4.2.2 An analytical method for optimization under inequality constraints

Consider the general optimization problem given in eqn (2.44) and assume that the set **S** defined by the constraints is compact.[5] In a general case, with possibly hundreds or thousands of variables, numerical methods must be applied. However, for problems involving fewer variables, it is sometimes possible to solve the constrained optimization problem by simply finding all points that *could* be minima, and checking which of these points are the actual minima. Consider first the interior points of **S**, i.e. the points in **S** that are not on the boundary ∂**S**. From the discussion in Section 2.3.1 it is evident that the local optima can be found only at stationary points or at points where the derivative does not exist. Next, consider the boundary ∂**S** of **S**. One can show that, if a point $\mathbf{x}^* \in \partial$**S** is a local optimum of the restriction of f to ∂**S**, then $\mathbf{x}^*$ is also a local optimum of f under the constraint $\mathbf{x} \in \partial$**S**. Thus, once all stationary points in the interior of **S** have been found *as well as* all possible optima on the boundary ∂**S**, the global optimum (or optima) can be found simply by determining the value of the function $f(\mathbf{x})$ at these points. The method is given in condensed form in Algorithm 2.5.

---

1. Find all the stationary points of $f(\mathbf{x})$ in the interior of **S**.
2. Find the stationary points of the restriction of $f(\mathbf{x})$ to the boundary ∂**S** of **S**, including any corner points.
3. Investigate the points thus found, one by one, to find the minima (or maxima).

Algorithm 2.5: *An analytical method for optimization of a function $f(\mathbf{x})$ under inequality constraints defining a compact set* **S**.

Example 2.7
Determine the global minimum of the function

$$f(x_1, x_2) = x_1 - x_1^2 - 2x_2^2 + x_2, \qquad (2.62)$$

on the set **S** defined in Fig. 2.10.

Solution Consider first the stationary points of f on **S**. Taking the partial derivatives of f, we find

$$\frac{\partial f}{\partial x_1} = 1 - 2x_1, \qquad (2.63)$$

$$\frac{\partial f}{\partial x_2} = -4x_2 + 1. \qquad (2.64)$$

---

[5] A subset $\mathbf{S} \subseteq \mathbf{R}^n$ is **compact** if it is closed (i.e. contains its boundary) and bounded (i.e. contained in a hypersphere of finite size). An example of a compact set is the unit disc $\mathbf{S} = \{(x_1, x_2): x_1^2 + x_2^2 \leq 1\}$.

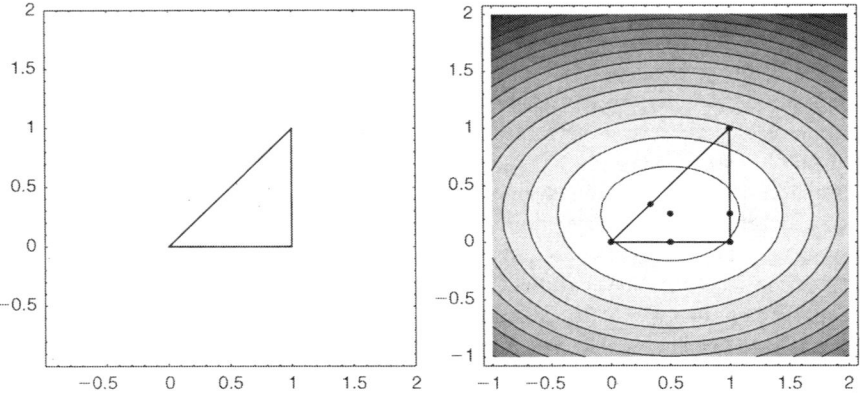

Figure 2.10: *Left panel: the set* **S** *considered in Example 2.7. The corners of the triangle are located at* $(0,0)^T$, $(1,0)^T$, *and* $(1,1)^T$. *Right panel: a contour plot of the function* $f(x_1,x_2) = x_1 - x_1^2 - 2x_2^2 + x_2$, *shown together with the points considered in the determination of the global minimum of f over* **S**. *Note that the contour plot clearly indicates that the global minimum is located at* $P_7 = (1,1)^T$.

Thus, there is only one stationary point, namely $P_1 = (1/2, 1/4)^T$. Since the partial derivatives of f exist everywhere on **S**, all that remains is to consider the boundary ∂**S**, which can be divided into three parts. Starting with the line $0 < x_1 < 1, x_2 = 0$, which defines an open set, we must determine the stationary points of $f(x_1, 0) = x_1 - x_1^2$. Taking the derivative, we find

$$1 - 2x_1 = 0, \tag{2.65}$$

and, therefore, $x_1 = 1/2$. Thus, another potential location of the global minimum is the point $P_2 = (1/2, 0)^T$. Continuing with the line $0 < x_2 < 1, x_1 = 1$, we next consider $f(1, x_2) = -2x_2^2 + x_2$, which, following the same procedure, yields a potential global minimum point $P_3 = (1, 1/4)^T$. Finally, the line $x_1 = x_2$ should be considered. Here $f(x_1, x_1) = 2x_1 - 3x_1^2$, and $P_4 = (1/3, 1/3)^T$ emerges as a point to consider. Finally, we must also consider the corners, i.e. $P_5 = (0,0)^T, P_6 = (1,0)^T$, and $P_7 = (1,1)^T$. Thus, there are seven points that are candidates for the global minimum and, determining the function value at all these points, one finds that the global minimum is actually located at one of the corner points, namely $P_7 = (1,1)^T$, for which f takes the value -1. ∎

2.4.2.3 Penalty methods

To conclude the presentation of classical optimization methods, we shall consider penalty methods, which are a special case of the interior point methods mentioned above, and which transform a constrained optimization problem of the kind given in eqn (2.44) to an unconstrained problem that can then be solved using standard numerical methods (such as gradient descent or Newton's method) for such problems. Consider the **penalty function**

$$p(\mathbf{x}; \mu) = \mu \left(\sum_{i=1}^{m} (\max\{g_i(\mathbf{x}), 0\})^2 + \sum_{i=1}^{k} (h_i(\mathbf{x}))^2 \right), \qquad (2.66)$$

where μ is a positive parameter. Inspection of $p(\mathbf{x}; \mu)$ shows that it takes non-negative values for all $\mathbf{x}$, and that $p(\mathbf{x}; \mu) = 0$ if, and only if, all constraints are satisfied. In fact, despite the max-function in the first sum, $p(\mathbf{x}, \mu)$ is continuously differentiable provided that all the constraint functions are. Thus, the problem of minimizing $f(\mathbf{x})$, subject to the constraints defined by $g_i(\mathbf{x})$ and $h_i(\mathbf{x})$ can be presented as the problem of minimizing (with respect to $\mathbf{x}$)

$$f_p(\mathbf{x}; \mu) = f(\mathbf{x}) + p(\mathbf{x}; \mu), \qquad (2.67)$$

without constraints. The **penalty method** proceeds by solving the unconstrained problem for a given value of μ, increasing μ (normally by a constant factor $\alpha > 1$), solving the unconstrained problem for the new value of μ, etc. As μ increases, the optima of $f_p(\mathbf{x}; \mu)$ will, under certain conditions that will not be given here, tend towards the optima of $f(\mathbf{x})$ under the constraints g_i and h_i, even though numerical problems may occur for very large values of μ. The method is described in Algorithm 2.6.

---

1. Starting from the objective function $f(\mathbf{x})$, form the function $f_p(\mathbf{x}; \mu) = f(\mathbf{x}) + p(\mathbf{x}; \mu)$, with $p(\mathbf{x}; \mu)$ given by eqn (2.66). Set μ to a small value $\mu_0 > 0$.
2. Solve the unconstrained optimization problem

   minimize $f_p(\mathbf{x}; \mu_j)$

   using a suitable method, e.g. gradient descent. $j = 0$ in the first iteration.
3. Form the next value of μ, e.g. as $\mu_{j+1} = \alpha \mu_j$, where α is a constant > 1.
4. Repeat steps 2 and 3. For large values of μ, the minimum of $f_p(\mathbf{x}; \mu_j)$ approaches the minimum of $f(\mathbf{x})$ subject to the constraints.

Algorithm 2.6: *The penalty method. Note that numerical problems may occur for large values of μ.*

Example 2.8
As a very simple example of the penalty method, consider the problem of finding the minimum of

$$f(x) = x^2 - x + 1, \qquad (2.68)$$

subject to the inequality constraint

$$x \geq 1. \qquad (2.69)$$

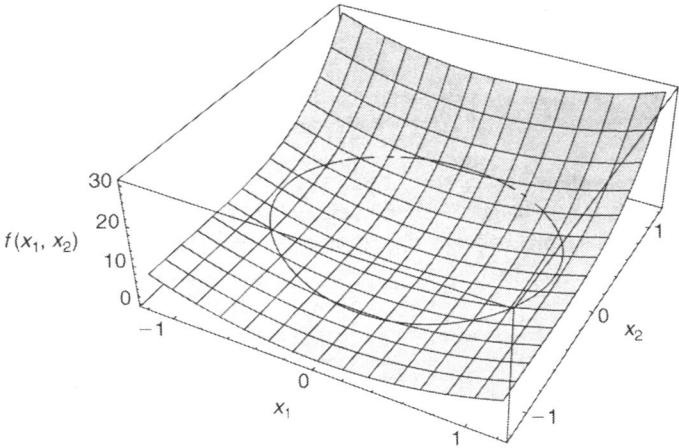

Figure 2.11: *The function $f(x_1, x_2) = 5x_1^2 + (x_2 + 1)^4$. The contour illustrates the equality constraint $h(x_1, x_2) = x_1^2 + 2x_2^2 - 1 = 0$.*

Solution It is obvious that the (global) minimum occurs at the boundary point $x^* = 1$. Let us now use the penalty method: following eqn (2.66), the penalty function takes the form

$$p(x) = \mu(\max\{0, (1-x)\})^2, \qquad (2.70)$$

Thus, we obtain

$$f_\mathrm{p}(x; \mu) = \begin{cases} x^2 - x + 1 + \mu(1-x)^2 & \text{if } x \leq 1, \\ x^2 - x + 1 & \text{otherwise.} \end{cases} \qquad (2.71)$$

Taking the derivative of $f_\mathrm{p}(x)$, it is easy to find that the stationary point ($f_\mathrm{p}'(x) = 0$) occurs at

$$x^*(\mu) = \frac{1 + 2\mu}{2 + 2\mu} \qquad (2.72)$$

Thus, as $\mu \to \infty$, $x^*(\mu) \to 1$. ∎

Example 2.9
Solve the minimization problem

$$\left.\begin{array}{l} \text{minimize } f(x_1, x_2) = 5x_1^2 + (x_2 + 1)^4 \\ \text{subject to } h(x_1, x_2) = x_1^2 + 2x_2^2 - 1 = 0 \end{array}\right\} \qquad (2.73)$$

The problem is illustrated in Fig. 2.11.

Solution The equality constraint leads to the penalty function

$$p(x_1, x_2) = \mu \left(x_1^2 + 2x_2^2 - 1\right)^2, \qquad (2.74)$$

so that

$$f_\mathrm{p}(x_1, x_2; \mu) = 5x_1^2 + (x_2 + 1)^4 + \mu \left(x_1^2 + 2x_2^2 - 1\right)^2. \qquad (2.75)$$

Table 2.1: *The minimum of $f_p(x_1, x_2; \mu)$ in Example 2.9, obtained for different values of μ. As μ increases, $\mathbf{x}^*(\mu)$ approaches the actual minimum of $f(x_1, x_2)$, subject to the constraint $h(x_1, x_2) = 0$, namely $\mathbf{x}^* = (0, -1/\sqrt{2})^T$.*

μ	$\mathbf{x}^*(\mu)$
0.000	$(0.000, -1.000)^T$
1.000	$(0.000, -0.718)^T$
10.00	$(0.000, -0.708)^T$
100.0	$(0.000, -0.707)^T$

The unconstrained problem of minimizing $f_p(x_1, x_2; \mu)$ can, for example, be solved using gradient descent or Newton's method. The solutions found for various values of μ are given in Table 2.1. As can be seen, for large values of μ, the minimum of $f_p(x_1, x_2; \mu)$ converges towards the minimum $\mathbf{x}^* = (0, -1/\sqrt{2})$ of $f(x_1, x_2)$, subject to the constraint $h(x_1, x_2) = 0$. ∎

2.5 Limitations of classical optimization

From the discussion above, it should be evident that classical optimization methods can be used for solving a wide variety of optimization problems. In particular, if an optimization problem is known to be convex, classical optimization is generally the best alternative. However, for some problems, other methods are more appropriate. First of all, while applicable to problems involving few variables, the analytical methods presented above soon become inadequate when problems with many variables are considered. Furthermore, even the numerical methods, such as gradient descent and Newton's method are inappropriate for some problems, for example those that involve non-differentiable objective functions. There are, however, classical optimization methods (not listed above) that can cope with non-differentiability, but even such methods fail to provide solutions in certain cases. Two important cases are (1) problems in which the objective function cannot be specified explicitly as a mathematical function, and (2) problems in which the number of variables itself varies during optimization. An example of the first case is the optimization of simple robotic control systems exemplified in Section 1.3, in which the values of the objective function were obtained as a result of a lengthy simulation procedure. As an example of the second case, one can consider, say, the optimization of a neural network (see Appendix A, Section A.2.3) in a problem where the network size cannot be specified *a priori*. In the remainder of this book, we shall consider optimization algorithms that, in addition to all the problems that can be solved by classical optimization, also can handle problems of the kinds just mentioned.

Exercises

2.1 Determine the stationary points of the function $f(x_1,x_2)=2x_1^3+x_1x_2^2+5x_1^2+x_2^2$.

2.2 Find the local optima of the function $f(x_1,x_2)=x_1^3+2x_1^2+x_1x_2+x_2^2$.

2.3 Consider the function $f(x_1,x_2)=2x_1^2+x_2^2-2x_1x_2$. Is this function convex?

2.4 Given two twice continuously differentiable functions of a single variable, $g(x)$ and $h(x)$, under what conditions is the composition $f(x)=g(h(x))$ convex?

2.5 Consider the function $f(x_1,x_2)=x_1^4+3x_1x_2-2x_1+2x_2^4$. Determine the direction of steepest descent at the point $P=(2,2)^T$. Next, determine the point reached after one step of the gradient descent method.

2.6 Use gradient descent (with a fixed step length η of your choice) to find a local minimum $\mathbf{x}^*$ of the function $f(x_1,x_2)=3x_1^4+3x_1^2x_2^2+x_1^2+2x_2^4$, starting from the point $P_0=(1,1)^T$. Prove that $\mathbf{x}^*$ is also the *global* minimum of f.

2.7 Find the gradient of the function

$$f(x_1,x_2) = x_1^4 + x_1x_2 + x_2^2 \tag{2.76}$$

at the point $(1,1)^T$. Next, starting from this point, take one step of gradient descent (including the line search needed to find the minimum along the search direction). Which point $(x_1^*, x_2^*)^T$ is reached after this step? Finally, demonstrate (numerically) that the search direction obtained in the next gradient descent step is perpendicular to the first search direction determined above.

2.8 Use Newton's method to find a local minimum x^* of the function $f(x)=x^4-x^3+x^2-x+1$, starting from the point $x=1$. Is x^* also the global minimum?

2.9 Use the method of Lagrange multipliers to determine the minimum value of the function $f(x_1,x_2)=-x_1x_2$, subject to the constraint $x_1^2+2x_2^2-1=0$.

2.10 Using the method of Lagrange multipliers, find the minimum value of the function

$$f(x_1,x_2) = x_1^2 x_2 \tag{2.77}$$

subject to the equality constraint

$$2x_1^6 + 3x_1^4x_2^2 + 8x_2^6 - 36 = 0. \tag{2.78}$$

2.11 Determine the minimum value of the function $f(x_1,x_2)=x_1^2x_2-3x_1x_2+2x_1-4x_2$ over the set $\mathbf{S}=\{(x_1,x_2):|x_1|\leq 2, 0\leq x_2\leq 2\}$.

2.12 Use the penalty method to find the minimum of the function $f(x_1,x_2)=x_1^2+2x_2^2$, subject to the constraint $2x_1^2+x_2=3$.

2.13 Determine the *maximum* value of the function

$$f(x_1,x_2) = 2x_1^2 - 4x_1 + x_2^2 + 2x_2, \tag{2.79}$$

subject to the constraint

$$2x_1^2 + x_2^2 \leq 12. \tag{2.80}$$

Chapter 3

Evolutionary algorithms

As the name implies, **evolutionary algorithms** (EAs) are based on processes similar to those that occur during biological evolution. Hence, most of the terminology regarding EAs has its origin in biology and, even though EAs are greatly simplified compared to the processes that occur in biological evolution, a basic understanding of the biological counterpart is highly relevant. Thus, we shall begin this chapter by a brief discussion of biological evolution, before proceeding to describe EAs, of which there are many different types. Towards the end of the chapter, in Section 3.5, we shall return to biology and discuss some of the simplifications (relative to the biological counterpart) introduced in EAs.

3.1 Biological background

Adaptation to challenges provided by the environment is a central process in nature, of which there are many examples, such as the shark skin discussed in Chapter 1. Other examples include the dynamics between predator and prey, where the both predators and prey gradually (over many generations) may develop, say, larger muscles allowing them to run faster, or more acute vision. Speaking of vision, the forward-pointing eyes of predators such as lions, and the sideways-pointing (or even, to some extent, backward-pointing) eyes of their prey are clear cases of adaptation. An extreme case of adaptation is **mimicry**, in which the outward appearance, or even the behaviour, of one species comes to resemble that of another species. Mimicry occurs for several reasons, one of the foremost is protection against predators. For example, the wings of an owl butterfly has two distinct, circular marks, the *ocelli*, resembling the eyes of an owl, which might serve the purpose of deterring avian predators whose own enemy is the owl. An alternative hypothesis is that the *ocelli* serve to deflect an attacker from the vital parts of the butterfly's body towards more peripheral parts, such as the wing tips. In either case, the *ocelli* have an adaptive value, and there are numerous other examples of mimicry throughout the animal kingdom.

Adaptation can be seen as a form of optimization, even though the optimization problem in nature differs significantly from the scientific and engineering problems normally associated with optimization: in nature, the target is always moving, in the sense that all species are subject to simultaneous evolution, and to concurrent changes in the environment. By contrast, in engineering problems, the desired goal is normally fixed and specified in advance. Nevertheless, in view of the amazing complexity of biological organisms, and the perplexing ingenuity of some of the designs found by evolution, there is sufficient motivation for developing optimization algorithms based on natural evolution. Clearly, in order to do so, one must have a basic understanding of the central concepts of biological evolution, a topic that will now be introduced briefly.

One of the central concepts in the theory of evolution is the notion of a **population**, that is, a group of individuals of the same **species**, i.e. individuals that can mate and have (fertile) offspring. **Darwin's theory of evolution** introduces the concept of gradual, hereditary change in the individuals of a species. In order for the properties of individuals to be transferred from one generation to the next, they must somehow be stored. The necessary tools for elucidating the process of storage and transfer of information in biological organisms, on a molecular level, were not available in Darwin's time. The discovery of DNA and the subsequent unravelling of the basic processes of natural evolution constitute one of the great successes of molecular biology during the last half-century or so [11].

How is the transferable information stored in an individual? The answer is that each individual of a species carries (in each cell of its body) a **genome** that, in higher animals, consists of *several* **chromosomes** in the form of DNA molecules. Each chromosome, in turn, contains a large number of **genes**, which are the units of heredity and which encode the information needed to build and maintain an individual. The number of chromosomes, and therefore also the number of genes, varies between species. In humans, each **somatic cell**[1] contains 23 pairs of chromosomes, and the total number of genes is estimated to be around 20,000–25,000 [47].

Each gene is composed of a sequence of **bases**. There are four molecules that serve as bases in chromosomes (or DNA molecules), and they are denoted A, C, G and T. Thus, the information is stored in a digital fashion, using an alphabet with four symbols, as illustrated in Fig. 3.1. The DNA molecules are **double-stranded** such that A always resides opposite T, and C always resides opposite G. This double-strandedness is crucial during replication of DNA (a process that precedes cell division). In replication, a DNA molecule is split, and each strand

---

[1] One distinguishes between somatic cells, which form the body of an organism, i.e. its bones, muscles, skin, etc., and **germ cells** (or sex cells), which are active in reproduction. Whereas somatic cells are **diploid**, meaning that they contain 23 *pairs* of chromosomes, germ cells are **haploid**, i.e. they only contain a single copy of each chromosome.

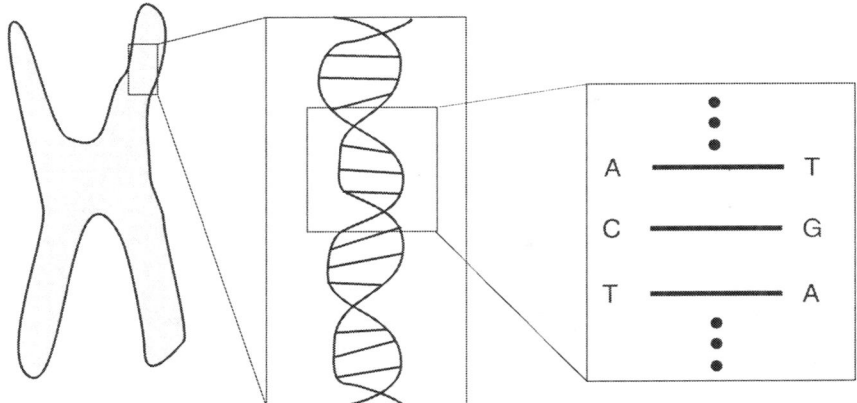

Figure 3.1: *A schematic representation of a chromosome is shown in the left side of the figure. The two blow-ups on the right show the individual base pairs. Note that A is always paired with T, and C is always paired with G.*

serves as a template for the creation of the opposite strand. During development, as well as during the life of an individual, the DNA is read by an **enzyme**[2] called **RNA polymerase**, and this process, known as **transcription**, produces another type of molecule called **messenger RNA** (mRNA). In this process, the DNA is (temporarily) split so that the RNA polymerase can access one sequence of bases. Then, during the formation of the mRNA molecule, the RNA polymerase moves along the DNA molecule, using it as a template to form a chain of bases, as shown in Fig. 3.2. Note that the T base is replaced by another base (U) in mRNA molecules.

Next, the **proteins**, which are chains of **amino acids**, are generated in a process called **translation** using mRNA as a template. A greatly simplified version of the translation process is given in Fig. 3.3. In translation, which takes place in the **ribosomes**, each sequence of three bases (referred to as a **codon**) codes for an amino acid, or for a *start* or *stop* command. Thus, for example, the sequence AUG is (normally) used as the start codon. Because there are only 20 amino acids, and $4^3 = 64$ possible three-letter combinations using the alphabet {A, C, G, U}, there is some redundancy in the code. For example, the codons GCU, GCC, GCA and GCG all code for the amino acid *alanine*. Other amino acids such as *tryptophan* are encoded only by one single sequence (UGG in the case of *tryptophan*). As shown in Fig. 3.3, the amino acids are transported to the ribosomes by **transfer RNA** (tRNA) molecules and are then used in the growing amino acid chain forming the protein.

The function and purpose of genes is thus to provide the information needed to form proteins, which are the building blocks of life and are involved in one way

---

[2] Enzymes, which catalyze chemical reactions in the body, are a special kind of protein.

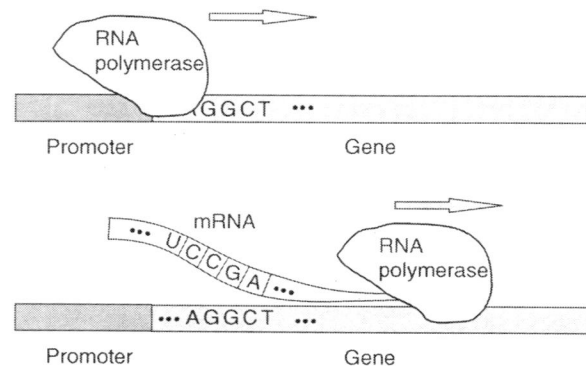

Figure 3.2: *The transcription process. The RNA polymerase binds to a promoter region and then moves along the DNA molecule, generating an mRNA molecule by joining bases available in the cell.*

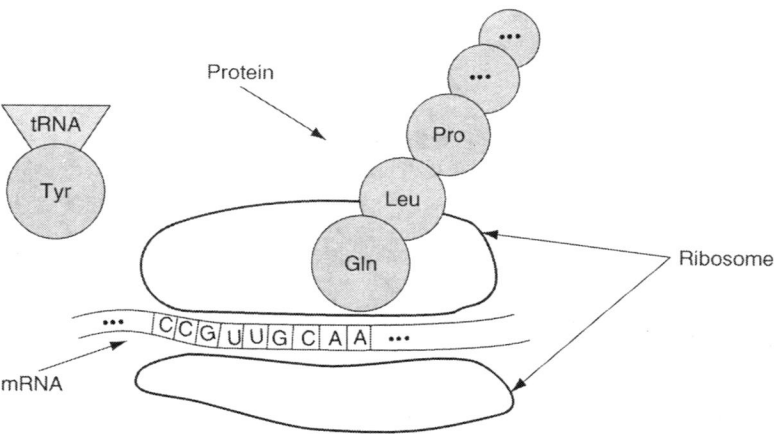

Figure 3.3: *The translation process. The tRNA molecules transport amino acids to the ribosomes. In the case shown in the figure, a tRNA molecule is just arriving, carrying a* tyrosine *amino acid. The growing chain of amino acids generated at the ribosomes will, when completed, form a protein. In this example, the amino acid* Glutamine (Gln), *encoded by the sequence CAA, has just been added to the protein. Note that the figure is greatly simplified; in reality, the process is quite complex.*

or another in almost every activity that takes place inside the living cell. However, not all genes are **expressed** (i.e. active) at all times. For example, some genes are primarily active during the development of the individual, whereas other genes are active during adult life. Also, different cells in a body may show different patterns of activity, even though the genetic material in each cell of a given individual is the same. Some genes, called **regulatory genes**, serve the purpose of regulating the activity of other genes, whereas other genes, called **structural genes**, are responsible for generating the proteins that make up the cells of the body, i.e. bones, muscle tissue, etc. The various functions of, and interactions between, genes will be further discussed in Section 3.5. Returning to the DNA molecules, it should be noted that each gene can have several settings, known as **alleles**, and that different alleles are responsible for the various aspects of an individual's appearance, such as eye color and hair color in humans. Now, the complete genome of an individual, with all its settings (encoding, e.g. hair color, eye color, etc.), is known as the **genotype**. During development, the stored information is decoded, resulting in an individual carrying the traits encoded in the genome. The individual, with all its traits, is known as the **phenotype**, corresponding to the genotype. It should be noted, however, that not all phenotypic traits are encoded as single genes. On the contrary, many traits require a complex interaction between several genes. Furthermore, not all genes encode something that is as easy to visualize as, e.g. eye color.

Two central concepts in evolution are **fitness** and **reproduction**, and these concepts are strongly intertwined. In fact, by some definitions, fitness equals the probability of reproducing. Individuals that are well adapted to their environment (which includes not only the climate and geography of the region where the individual lives, but also other members of the same species as well as members of other species), those that are stronger or more intelligent than the others, have a larger chance to reproduce, and thus to spread their genetic material, resulting in more individuals having these properties. Thus, over long periods of time, the properties of biological organisms will gradually change, and it is this process that forms the basis for the evolutionary algorithms considered below.

Reproduction is the central moment for evolutionary change, and two different kinds are found in nature: **asexual reproduction**, involving a single individual, and **sexual reproduction**, involving two individuals. Asexual reproduction occurs in single-celled organisms such as bacteria. In bacteria, the (single) chromosome is closed on itself, in a coil-like structure. During reproduction, a copy of the chromosome is made and the cell then splits into two. Sexual reproduction is a much more complex procedure, which will not be described in detail here. Suffice it to say that, during this process, the genetic materials of two individuals are combined, some genes being taken from one parent and others from the other parent. The copying of genetic information takes place with remarkable accuracy, but nevertheless there occur some errors. These errors are known as **mutations**, and constitute the providers of new information for evolution to act upon.

The description above summarizes some of the main concepts of evolutionary and molecular biology, albeit in a simplified, and somewhat incomplete, fashion. A more thorough presentation of these topics can be found in Ref. [43]. We will now proceed to introduce EAs, i.e. stochastic optimization algorithms inspired by natural evolution. EAs is an umbrella term that encompasses a variety of different algorithms. Here, we shall only consider three types of EAs, namely genetic algorithms (GAs), linear genetic programming (LGP) and interactive evolutionary computation (IEC).

3.2 Genetic algorithms

We shall start our study of EAs by describing a basic GA as well as its practical use in a simple function optimization example. Next, the various operators will be discussed in some detail. Consider the problem of finding the *maximum* of a function $f(x_1, \ldots, x_n)$ of n variables. The set of allowed values for the variables $\mathbf{x} = (x_1, \ldots, x_n)^T$ is referred to as the **search space**. Now, in order to apply a GA to solve this problem, the variables must be encoded in strings of digits referred to as **chromosomes**. The digits constituting the chromosome are referred to as **genes**, in accordance with the biological terminology introduced above. Thus, the genes encode the information stored in the chromosome, and there exists different **encoding schemes**. In the original GAs, introduced by Holland [32] and others in the 1970s, a **binary encoding scheme**[3] was employed in which the genes take the values 0 or 1.

When the algorithm is initialized, a **population** (i.e. a set) of N chromosomes c_i, $i = 1, \ldots, N$ is generated by assigning random values, normally with equal probability for the two alleles[4] 0 and 1, to the genes. The chromosomes thus formed constitute the first **generation**. Note that in the biological counterpart a population is defined as a set of individuals, rather than a set of chromosomes. However, in GAs the distinction between the chromosomes and the corresponding individuals is less relevant than in biological organisms, because of the simple mapping (illustrated in Example 3.1) that generates individuals from chromosomes. The complex developmental programmes seen in biological organisms are largely absent in GAs, even though exceptions exist [4, 19].

After initialization, each of the N chromosomes is decoded to form the corresponding individual, in this case consisting of the n variables[5] x_j, $j = 1, \ldots, n$. The procedure for obtaining the variables from the chromosomes can be implemented in various ways. One of the simpler ways is to divide the chromosome (of length m) into n equal parts consisting of $k = m/n$ bits, and considering each such part as a binary number which can then be converted to a decimal number representing the variable in question, as described in Example 3.1. Clearly, the accuracy of a variable is determined by the number of bits used.

---

[3] The main motivation for using binary encoding is, in fact, historical and, to some extent, aesthetic: the theoretical analysis of GAs becomes more compact and clear if a binary alphabet is used. However, in applications of GAs, other encoding methods such as real-number encoding described in Section 3.2.1 work just as well as binary encoding methods.

[4] In connection with GAs, the terms *allele* and *value* (of a gene) will be used interchangeably.

[5] When referring to the generic variables of a problem, the notation x_j, with a single subscript, will be used in order to conform with the notation used in Chapter 2. However, when referring to the variables obtained for individual i, two indices will be used, and the variables will then be written x_{ij}, where the first index enumerates the individuals, from 1 to N, and the second index enumerates the actual variables, from 1 to n.

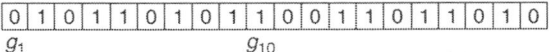

Figure 3.4: *A chromosome encoding two variables, x_1 and x_2, with 10-bit accuracy.*

Example 3.1
Consider the simple problem of finding, using a GA, the maximum of some function $f(x_1, x_2)$ of two variables that are to be encoded using 10 bits each. The search range, i.e. the allowed values for each variable, is in this example taken as $[-3, 3]$. At initialization, N strings of $m = 2 \times 10 = 20$ bits are thus formed randomly. In the decoding step for each chromosome, the first 10 bits are used to form x_1 and the remaining bits form x_2 (see also Fig. 3.4). When forming x_1, the first 10 bits of the chromosome are converted to a decimal number in the range $[0, 1]$ according to

$$x_{1,\text{tmp}} = \sum_{j=1}^{10} 2^{-j} g_j. \tag{3.1}$$

This value is then scaled to the required range using the transformation

$$x_1 = -3 + \frac{2 \times 3}{1 - 2^{-10}} x_{1,\text{tmp}}. \tag{3.2}$$

The second variable is obtained using the last 10 genes

$$x_{2,\text{tmp}} = \sum_{j=1}^{10} 2^{-j} g_{j+10} \tag{3.3}$$

and, scaling the result in the same way as for x_1, one obtains:

$$x_2 = -3 + \frac{2 \times 3}{1 - 2^{-10}} x_{2,\text{tmp}}. \tag{3.4}$$

In the particular case shown in the figure, one gets

$$x_{1,\text{tmp}} = 2^{-2} + 2^{-4} + 2^{-5} + 2^{-7} + 2^{-9} + 2^{-10} = \frac{363}{1024} \approx 0.3545 \tag{3.5}$$

so that

$$x_1 = -3 + 6 \times \frac{1024}{1023} \times \frac{363}{1024} = -\frac{27}{31} \approx -0.8710. \tag{3.6}$$

x_2 can be obtained in a similar way. ∎

Once the variables have been obtained, the individual is evaluated, a procedure that differs from problem to problem. The aim of the evaluation is to assign a **fitness value**, which can later be used when selecting individuals for reproduction. The fitness of individual i is denoted F_i. Normally, in GAs as in the biological counterpart, the fitness value is a goodness measure rather than an error measure, i.e. the aim is to *maximize* fitness. However, this does not, of course, exclude

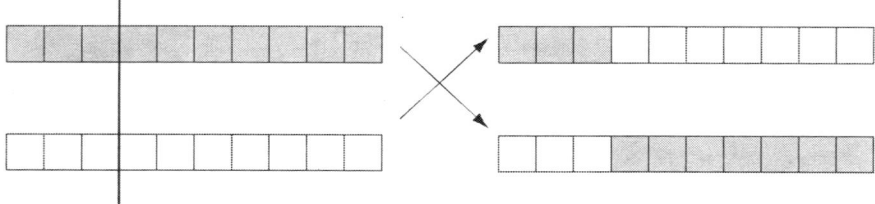

Figure 3.5: *The standard crossover procedure used in GAs. Each square corresponds to a gene. The crossover point, which is indicated by a thick line, is chosen randomly.*

minimization: minimizing a given measure is equivalent to maximizing its inverse. In the case of function maximization the evaluation is simple: the function value itself can be taken as the fitness value.

The procedure of decoding the chromosome, evaluating the corresponding individual and assigning a fitness measure is repeated until all N individuals have been evaluated. The next step is to form the second generation. Returning briefly to the biological counterpart, we can list the steps required to do so. First of all, there must be a process of **selection** in which the most fit individuals are selected as progenitors. However, we may also appreciate that the selection procedure should not be fully deterministic, i.e. it should not *always* favour the most fit individuals. Such a greedy procedure would easily lead to stagnation, especially in view of the fact that the initial population is generated randomly: an individual that happens to be better than the others, but still far from the global optimum may, if deterministic selection is applied, come to dominate the population, thus preventing the GA from finding the global optimum. In some cases, a less fit individual may contain a sequence of genes (or, to be strict, alleles) that will generate a highly fit individual when combined with genetic material from another individual, assuming that sexual reproduction is applied. Thus, the selection process is normally stochastic. A common approach is to select individuals (from the entire population) in direct proportion to their fitness, even though other methods exist as well, as we shall see below.

After selection, new individuals are formed through **reproduction**. In sexual reproduction, the genetic material of two individuals is combined. Individuals are therefore selected in pairs, using the selection procedure introduced above. In a standard GA, the genetic material is contained in a single chromosome of given length, with only one strand (unlike the double-stranded DNA molecules in biological organisms). Thus, a simple procedure for combining the genetic material of two individuals, a process referred to as **crossover**, consists of cutting the chromosomes at a randomly selected **crossover point** and then assembling the first part of the first chromosome with the second part of the second chromosome, and vice versa, as illustrated in Fig. 3.5.

The next step in the formation of new individuals is mutation. Of course, in a computer the copying errors referred to as mutations in biological organisms can

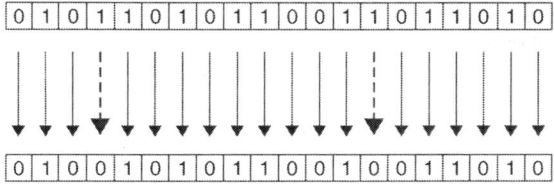

Figure 3.6: *The mutation procedure. Each gene is mutated with a small probability p_{mut}. In this case two mutations, marked with dashed arrows, occurred.*

easily be avoided. However, these errors play a crucial role in providing evolution with new material to work with. Thus, in GAs, once the new chromosomes have been generated through crossover, they are subjected to mutations in the form of random variation (bit flipping) of some, randomly selected, genes. Typically, mutations are carried out on a gene-by-gene basis in which the probability of mutation of any given gene equals a pre-specified **mutation probability** p_{mut}. In practice, this is done by generating, for each gene, a random number $r \in [0, 1]$, and mutating the gene if $r < p_{\text{mut}}$. The procedure is illustrated in Fig. 3.6.

The final step is **replacement**: after selection, crossover and mutation, there are now two populations available, each with N individuals. In **generational replacement**, the individuals of the first generation are all discarded, and the N new individuals thus form the second generation, which is evaluated in the same way as the first generation. The procedure is then repeated, generation after generation, until a satisfactory solution to the problem has been found. The basic GA just described is summarized in Algorithm 3.1. Before describing the various genetic operators in detail, we shall briefly consider a specific example.

Example 3.2
Use a GA to find the maximum of the function

$$f(x_1, x_2) = e^{-x_1^2 - x_2^2}, \tag{3.7}$$

in the range $x_1, x_2 \in [-2, 2]$. The function is illustrated in the top left panel of Fig. 3.7.

Solution It is elementary to see that, in the given interval, the function reaches its maximum ($=1$) at $x_1 = x_2 = 0$. In this simple case, one obviously does not need to apply a sophisticated optimization method, but the problem is appropriate for illustrating the basic operation of a GA, particularly because a function of two variables can easily be visualized. Assuming that a binary encoding scheme is used, the first step is to select the accuracy, i.e. the number of bits per variable (k) in the chromosomes. The overhead involved in extending the chromosome length a bit is almost negligible, so in a purely mathematical problem such as this one, where there is no measurement error, one might as well choose a rather large k. Let us set $k = 25$ so that the smallest allowed increment (see also eqn (3.9) below) equals

$$\frac{4}{1 - 2^{-k}} 2^{-k} \approx 1.19 \times 10^{-7}. \tag{3.8}$$

> 1. Initialize the population by randomly generating N binary strings (chromosomes) $c_i, i = 1, \ldots, N$ of length $m = kn$, where k denotes the number of bits per variable.
> 2. Evaluate the individuals:
>
>   2.1. Decode chromosome c_i to form the corresponding variables $x_{ij}, j = 1, \ldots, n$ (or, in vector form, $\mathbf{x}_i$).
>   2.2. Evaluate the objective function f using the variable values obtained in the previous step, and assign a fitness value $F_i = f(\mathbf{x}_i)$.
>   2.3. Repeat steps 2.1 and 2.2 until the entire population has been evaluated.
>
> 3. Form the next generation:
>
>   3.1. Select two individuals i_1 and i_2 from the evaluated population, such that individuals with high fitness have a greater probability of being selected than individuals with low fitness.
>   3.2. Generate two new chromosomes by crossing the two selected chromosomes c_{i_1} and c_{i_2}.
>   3.3. Mutate the two chromosomes generated in the previous step.
>   3.4. Repeat steps 3.1–3.3 until N new individuals have been generated. Then replace the N old individuals by the N newly generated individuals.
>
> 4. Return to step 2, unless the termination criterion has been reached.

Algorithm 3.1: *Basic genetic algorithm. See the main text for a complete description of the algorithm.*

The initial step of the GA thus comprises generating random strings consisting of $m = 2 \times 25 = 50$ bits. The choice of the population size N normally affects the results. Typical values of the population size range from around 30 to 1,000. For this simple problem, however, we shall choose $N = 10$.

The initial population is shown in the top right panel of Fig. 3.7. Evaluating the 10 individuals, it was found, in the run illustrated in the figure, that the best individual of the initial generation had a fitness of around 0.5789. Next, the second generation was formed. Selection was carried out stochastically in direct proportion to fitness (using the roulette-wheel method described in Section 3.2.1 below). The selection procedure was carried out 10 times, with replacement, meaning that individuals could be selected more than once. From each of the five pairs thus generated, new chromosomes were formed using the crossover procedure shown in Fig. 3.5, with randomly selected crossover points. Next, the newly formed chromosomes were mutated, as shown in Fig. 3.6, with the mutation rate $p_{\text{mut}} = 0.02$ so that on average, one bit per chromosome was changed. The best individual in the second generation obtained a fitness of around 0.8552. The progress of the GA in a typical run is shown in the two lower panels of Fig. 3.7. In this run, a value of $f(x_1, x_2) \approx 0.9987$ was obtained after 25 generations, i.e. after the evaluation of 250 individuals.

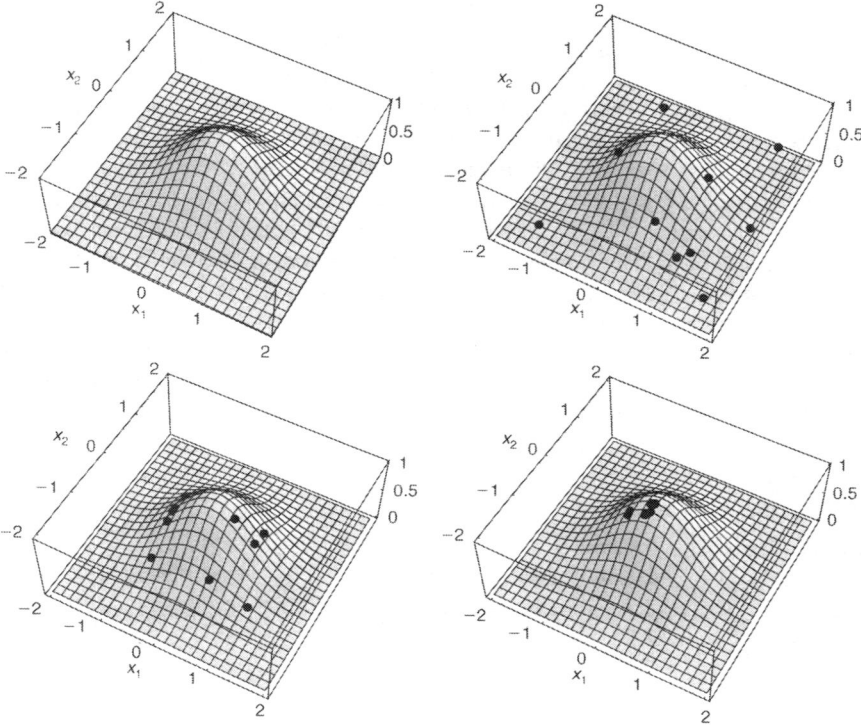

Figure 3.7: *The progress of a GA searching for the maximum of the function* $f(x_1, x_2) = e^{-x_1^2 - x_2^2}$. *The upper right panel shows the initial population, whereas the lower left and lower right panels show the population after 2 and 25 generations, respectively.*

Note that the results shown in Fig. 3.7 are taken from a single run. However, because of the stochastic nature of the GA, different runs take different paths towards the optimum, a fact that can be illustrated by letting the GA run many times, starting from different random initial populations. Such an analysis was carried out, repeating the run described above 100 times, in each run evaluating 250 individuals. The result was an average best fitness in the final generation (over the 100 runs) of 0.9810, with a span from a worst result of 0.7522 to a best result of 0.9999. The median of the best fitness in the final generation was 0.9935.

It should be noted that these results are an underestimate of the performance of the GA, because there is no guarantee, following Algorithm 3.1, that the best individual of a given generation is not eliminated during selection or as a result of crossover or mutation. Thus, the results in the final generation are not necessarily the best results obtained. For example, in the run depicted in Fig. 3.7, an individual with fitness 0.9995 was found in generation 22, but it was eliminated in the subsequent selection stage. In order to avoid losing good individuals, the best individual of any given generation is normally copied unchanged to the next generation, a procedure called elitism that will be studied towards the end of Section 3.2.1. ∎

3.2.1 Components of genetic algorithms

As illustrated above, GAs involve several steps, and each component or operator[6] (e.g. selection, crossover, etc.) can be implemented in various ways. In addition, most components are associated with one or more parameters that, of course, must be set to appropriate values. In this section, we shall consider the components and parameters of GAs, giving a few alternative implementations for each component. However, just knowing of the existence of different choices of components is not sufficient: one must also be able to *choose* wisely among the available options, and to set the corresponding parameters correctly. Alas, there is no single choice of components and parameters that does better than other choices over all problems. However, by considering the performance of different GAs on benchmark problems, one may at least draw some conclusions regarding the performance of various (combinations of) components and the appropriate range of their parameters. A set of benchmark functions, some of which will be used in the examples below, is given in Appendix D.

3.2.1.1 Encoding schemes
In Example 3.1, standard binary encoding was used, in which the allowed alleles are 0 and 1, and where a generic variable x is formed from genes $g_1, \ldots, g_k$ as

$$x = -d + \frac{2d}{1 - 2^{-k}}(2^{-1}g_1 + \cdots + 2^{-k}g_k), \qquad (3.9)$$

giving a value in the range $[-d, d]$. Binary encoding schemes were employed in the original GAs introduced by Holland [32] and are still frequently used. However, even though a binary representation simplifies the analytical treatment of GAs (see Appendix B, Section B.2), in practical applications the decoding step introduced in eqn (3.9) may be perceived as unnecessary, and it can be avoided by using **real-number encoding** in which each gene g holds a floating-point number in the range $[0, 1]$ from which the corresponding variable x is simply obtained as[7]

$$x = -d + 2dg, \qquad (3.10)$$

again resulting in a value in the range $[-d, d]$.

---

[6] The terms *component* and *operator* will be used interchangeably in connection with GAs.

[7] It is, of course, possible to let the range of the genes be $[-d, d]$ so that even the rescaling in eqn (3.10) becomes unnecessary. However, in some problems, different genes should have different ranges. Thus, by using the range $[0, 1]$ throughout the chromosome, and then rescaling according to eqn (3.10), only the decoding procedure of the GA must be adapted to the problem at hand, whereas, for example, the initialization and mutation procedures can be applied without any particular adaptation.

While there is no systematic difference in the performance of GAs as a result of choosing either binary or real-number encoding, it should be noted that in real-number encoding each gene carries more information than in binary encoding. Therefore, a random change in one gene in a binary chromosome generally leads to a smaller change in the corresponding variable x than if real-number encoding is used. For this reason, specialized mutation operators (creep mutations), which are discussed later in this section, are commonly used in connection with real-number encoding schemes. One should also note that the number of potential crossover points is reduced significantly if real-number encoding is used. Thus, if the problem under consideration involves only a small number of variables (2–3, say), the positive effects of crossover may be strongly diminished if real-number encoding is used. However, because GAs are typically applied in problems involving many variables, this problem rarely occurs in practice.

Returning to binary encoding schemes, we note that a small change in a variable x may require flipping many bits which, in turn, is an unlikely event. Thus, the algorithm may get stuck simply as a result of the encoding scheme. Consider, as an example, a 10-bit binary encoding scheme, applied in a case with a single variable x, and assume that the best possible chromosome is 1000000000. Now, if the population happens to converge to 0111111111, the decoded value x will be near the optimal one, but taking the final step to the very best chromosome will require flipping all 10 bits in the chromosome. An alternative representation which avoids this problem is the **Gray code** [25]. A Gray code is simply a binary representation of the integers in the range $[0, 2^k - 1]$ such that going from an integer i to $i+1$ requires only 1 bit to change in the representation. Letting $\gamma(g_1, \ldots, g_k)$ denote the integer i obtained from a set of genes $g_1, \ldots, g_k$, the corresponding variable x (for use in a GA) can be obtained as

$$x = -d + \frac{2d}{1 - 2^{-k}} 2^{-k} \gamma(g_1, \ldots, g_k). \quad (3.11)$$

A Gray code representation of the numbers 0, 1, 2, 3 is given by $\gamma(00) = 0$, $\gamma(01) = 1$, $\gamma(11) = 2$, $\gamma(10) = 3$. Clearly, several different Gray code representations can be generated for the 2-bit case (another example is $10 \leftrightarrow 0$, $11 \leftrightarrow 1$, $01 \leftrightarrow 2$, $00 \leftrightarrow 3$). However, those representations differ from the original one only in that the binary numbers have been permuted or inverted. An interesting question is whether the Gray code is unique if permutations and inversions are disregarded. The answer turns out to be negative for $k > 3$ (see Exercise 3.3).

A further alternative to the standard binary encoding scheme is **messy encoding**, which generates a less position-dependent representation by associating each gene with a number determining its position in the chromosome. Thus, a messy chromosome of length m is represented as

$$c = ((p_1, g_{p_1}), (p_2, g_{p_2}), \ldots, (p_j, g_{p_j}), \ldots, (p_m, g_{p_m})), \quad (3.12)$$

where p_j denotes the position in the chromosome, and g_{p_j} the corresponding allele. Thus, for example, the messy chromosome ((1,0),(5,1),(3,0),(4,1),(2,1)) represents the chromosome 01011. When messy encoding is used, the initialization procedure consists of generating, for each messy chromosome, a random permutation of the numbers 1, ..., m to be used as the position markers p_j as well as a sequence of random bits to be used as the alleles g_{p_j}. Two obvious problems that may occur in connection with messy encoding schemes are missing positions and duplicates. In the latter case, a messy chromosome may (as a result of crossover) contain several copies of a given gene position p_j, e.g. ((1,0),(3,1),(3,0),(2,1),...). One way of resolving such conflicts is simply to use the first occurrence of a given gene position and thus discard all other occurrences. The case of missing gene positions can be mitigated simply by extending the length of the messy chromosomes beyond the required length m, thus reducing (but not eliminating) the probability of missing gene positions as a result of crossover. If some positions are still missing, one may, for example, set those missing positions to a given value, i.e. either 0 or 1.

The encoding schemes encountered thus far are all applicable in problems involving, say, function optimization, where the chromosomes are required to generate a vector $\mathbf{x} = (x_1, \ldots, x_n)^T$ of real-valued variables. In addition, other encoding schemes, tailored to particular problems, exist as well. An example is **permutation encoding** in which each chromosome holds a permutation (i.e. an ordering) of the integers $1, \ldots, n$. Permutation encoding is applicable, for example, in the travelling salesman problem (TSP) considered in Chapter 4.

3.2.1.2 Selection

Selection, the process of choosing the chromosomes that will be used when forming new individuals, can be carried out in many different ways; the two most common are **roulette-wheel selection** and **tournament selection**. In roulette-wheel selection, individuals are selected from the population using a fitness-proportional procedure, which can be visualized as a roulette-wheel (hence the name) such that each individual occupies a slice proportional to its fitness. When this imaginary wheel is turned and eventually comes to a halt, the probability of an individual being selected will be proportional to its fitness. The roulette wheel is, of course, just a metaphor. In practice, the procedure is carried out by forming the cumulative relative fitness values ϕ_j, defined as

$$\phi_j = \frac{\sum_{i=1}^{j} F_i}{\sum_{i=1}^{N} F_i}, \quad j = 1, \ldots, N, \tag{3.13}$$

where F_i denotes the fitness of individual i. Next, a random number $r \in [0, 1]$ is drawn, and the selected individual is taken as the one with the smallest j that satisfies

$$\phi_j > r. \tag{3.14}$$

The procedure is illustrated in Example 3.3.

EVOLUTIONARY ALGORITHMS 49

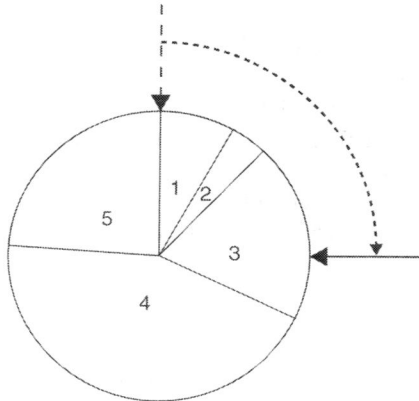

Figure 3.8: *Roulette-wheel selection in a population consisting of five individuals, described in detail in Example 3.3.*

Example 3.3
A roulette-wheel selection step is to be carried out on a population consisting of $N = 5$ individuals, with fitness values $F_1 = 2.0$, $F_2 = 1.0$, $F_3 = 5.0$, $F_4 = 11.0$ and $F_5 = 6.0$. The random number $r = 0.25$ is drawn. Which individual is selected?

Solution The roulette-wheel is illustrated in Fig. 3.8. The random number $r = 0.25$ corresponds to a rotation of 90 degrees (i.e. a quarter of a circle) of the wheel or, equivalently, rotation of the stopping point (indicated by an arrow in the figure) by 90 degrees. In this case, the arrow points to individual 3, which is thus selected. Mathematically, one would form the sum $\sum_{i=1}^{5} F_i = 25$ and then compute ϕ_j according to eqn (3.13). In this case (check!), $\phi_1 = 0.08$, $\phi_2 = 0.08 + 0.04 = 0.12$, $\phi_3 = 0.32$, $\phi_4 = 0.76$, and $\phi_5 = 1.00$. Thus, the smallest j for which $\phi_j > r = 0.25$ is $j = 3$, resulting in the selection of individual 3. ∎

Evidently, roulette-wheel selection is not very plausible from a biological point of view. An alternative procedure that better resembles the typical contests between pairs of animals (usually males) so often seen in nature is tournament selection. In its simplest form, tournament selection consists of picking two individuals randomly (i.e. with equal probability for all individuals, and, for simplicity, with replacement[8]) from the population, and then selecting the best individual of the pair, i.e. the one with higher fitness, with probability p_{tour}. Thus, the worse individual is selected with probability $1 - p_{tour}$. p_{tour} is referred to as the **tournament selection**

---

[8] If one wishes to eliminate the possibility of picking a pair consisting of two copies of the same individual, one may simply avoid using replacement, i.e. pick the second individual of the pair from the remaining $N - 1$ individuals. However, the probability of picking the same individual twice equals $1/N^2$ and, therefore, doing so has no systematic adverse effect on the progress of a typical GA run where N is of order 10^2 or larger.

parameter, and it typically takes values around 0.7–0.8. In practice, tournament selection is achieved by drawing a random number r in [0, 1], and selecting the better of the two individuals if $r < p_{\text{tour}}$. Otherwise, the worse individual is selected. Tournament selection can be generalized to involve more than two individuals. In the general case, j individuals are selected randomly from the population and the best individual is selected with probability p_{tour}. If this individual is not selected, the next step is to repeat the procedure for the remaining $j - 1$ individuals, again with probability p_{tour} of selecting the best individual. j is referred to as the **tournament size**. Example 3.4 illustrates the selection of a pair of individuals using tournament selection.

Example 3.4
Consider again the population of five individuals introduced in Example 3.3. An individual is to be selected using tournament selection with tournament size equal to two, and $p_{\text{tour}} = 0.8$. What is the probability p_1 of selecting individual 1 with fitness $F_1 = 2.0$?

Solution   Because the selection takes place with replacement, all pairs of individuals may occur, with equal probability, in this case equal to $1/(5 \times 5) = 1/25$. The pairs can be represented on a grid, as shown in Fig. 3.9. Now, for any given pair $(1, k)$ or $(k, 1)$ involving individual 1 (shown as shaded grid cells in the figure), the probability of selecting that individual equals p_{tour} if $F_1 > F_k$ and $1 - p_{\text{tour}}$ otherwise, except for the pair $(1, 1)$ consisting of two copies of individual 1, for which selection of that individual obviously occurs with probability 1. Thus, with the fitness values F_i given in Example 3.3, one obtains

$$p_1 = \frac{1}{25}(1 + 2p_{\text{tour}} + 6(1 - p_{\text{tour}})) = \frac{1}{25}(7 - 4p_{\text{tour}}) = 0.152. \quad (3.15)$$

∎

It should be noted that in tournament selection negative fitness values are allowed in principle, whereas in roulette-wheel selection all fitness values must be non-negative. Furthermore, in tournament selection it is only the relative fitness values that matter, not their absolute difference. This is not the case with roulette-wheel selection and, in fact, a common problem is that an individual which happens to have above-average fitness values in the first, randomly generated generation comes

(1, 1)	(1, 2)	(1, 3)	(1, 4)	(1, 5)
(2, 1)	(2, 2)	(2, 3)	(2, 4)	(2, 5)
(3, 1)	(3, 2)	(3, 3)	(3, 4)	(3, 5)
(4, 1)	(4, 2)	(4, 3)	(4, 4)	(4, 5)
(5, 1)	(5, 2)	(5, 3)	(5, 4)	(5, 5)

Figure 3.9: *Tournament selection, with tournament size equal to two, in a population consisting of five individuals, described in detail in Example 3.4.*

to dominate the population, which, in turn, sometimes leads to the GA getting stuck at a local optimum. This problem can, to some extent, be avoided by using **fitness ranking** which is a procedure for reassigning fitness values. In its simplest form, the best individual is given fitness N (i.e. equal to the population size), the second best individual is given fitness $N - 1$, and so on, down to the worst individual that is given fitness 1. Letting $R(i)$ denote the ranking of individual i, defined such that the best individual i_{best} has ranking $R(i_{\text{best}}) = 1$, the fitness values obtained by ranking can thus be summarized as

$$F_i^{\text{rank}} = (N + 1 - R(i)). \tag{3.16}$$

The ranking procedure can be generalized as

$$F_i^{\text{rank}} = F_{\text{max}} - (F_{\text{max}} - F_{\text{min}}) \left(\frac{R(i) - 1}{N - 1} \right), \tag{3.17}$$

which yields equidistant fitness values in the range $[F_{\text{min}}, F_{\text{max}}]$. The fitness values obtained before ranking, i.e. F_i, are referred to as **raw fitness values**, and we note that ranking removes the requirement that (raw) fitness values should be non-negative in connection with roulette-wheel selection.

The selection methods introduced thus far are the two most common ones. However, they suffer from one potential drawback: unless the selection parameters are gradually adjusted,[9] the extent to which high-fitness individuals are preferred over low-fitness ones (the **selection pressure**) is constant over time, something that might not always be favourable. In the early stages of a GA run, none of the individuals are likely to be very good compared to the global optimum, and the selection procedure would do well to include a natural, easily parameterized, mechanism for exploring the search space more or less freely in the early stages, without paying too much attention to fitness values, and then towards the end of a run resort to making small adjustments of the current best individuals. Thus, to summarize, the two selection methods just presented do not address the dilemma of exploration versus exploitation.

To mitigate this, one may use **Boltzmann selection**, a procedure that introduces concepts from physics into the mechanisms of GAs. In this selection scheme, the notion of a temperature T is introduced, and the basic idea is to use T as a tunable parameter for determining the selection pressure. The method derives its name from the fact that the corresponding equations (see below) are similar to the Boltzmann distribution which, among other things, can be used for determining the distribution of particle speeds in a gas. Boltzmann selection can be implemented in different ways that correspond either to roulette-wheel selection or tournament selection. An

---

[9] Specifically, the selection pressure exerted by tournament selection can be adjusted during a GA run by increasing either the tournament size or the tournament selection parameter p_{tour} (or both).

example of the former is to select individual i with probability p_i given by

$$p_i = \frac{e^{\frac{F_i}{T}}}{\sum_{j=1}^{N} e^{\frac{F_j}{T}}}, \qquad (3.18)$$

where, as usual, F_i denotes the fitness of individual i. Eqn (3.18) shows that individuals are selected with approximately equal probabilities if T is large, whereas for small T, individuals with high fitness are more likely to be selected. A Boltzmann selection scheme that resembles tournament selection first picks two individuals j and k randomly from the population. Next, the function

$$b(F_j, F_k) = \frac{1}{1 + e^{\frac{1}{T}\left(\frac{1}{F_j} - \frac{1}{F_k}\right)}} \qquad (3.19)$$

is considered, where F_j and F_k are the fitness values of the two individuals in the pair. During selection, a random number r is generated and the selected individual i is determined according to

$$i = \begin{cases} j & \text{if } b(F_j, F_k) > r, \\ k & \text{otherwise.} \end{cases} \qquad (3.20)$$

If T is large, selection occurs with almost equal probability for both individuals regardless of their fitness values. However, if T is small, the better of the two individuals is selected with a probability that approaches one as T tends to zero.

Thus, by starting from large values of T and then gradually lowering T towards zero, the GA will first carry out a more or less aimless exploration, and then as T is reduced, focus on exploitation of the results found. On the other hand, the introduction of Boltzmann selection raises the issues of how to set the initial value of T and how to reduce T during optimization, a question to which there is no definitive answer. Typically, the variation of T must be set through experimentation (see Exercise 3.4).

Selection normally occurs in a pairwise manner, i.e. the selection procedure is carried out twice, in order to obtain two individuals (or, rather, their chromosomes) that can then be subjected to crossover and mutation. As a final point, it should be noted that selection is done *with replacement*, i.e. the selected individual is *not* removed from the pool of available individuals. Thus, it is possible for a single individual to be the progenitor of several new individuals.

3.2.1.3 Crossover

Crossover is an essential component in GAs. It allows partial solutions from different regions of the search space to be assembled, thus allowing for a wide-ranging, non-local search. However, crossover may not always be beneficial: in GAs, typical population sizes (N) are around 30–1,000, much smaller than the thousands or millions of individuals found in many (though not all) biological populations. Through crossover, a successful individual will spread very quickly in the population, hence reducing diversity, a process akin to inbreeding in natural populations. Thus, crossover is, in fact, a bit *too* efficient and may cause the population to become trapped at a local optimum. It is therefore common to carry out crossover only with a certain probability p_c, referred to as the **crossover probability**. In cases where crossover is *not* carried out, the two selected individuals are simply copied as they are.

The crossover procedure can be implemented in various ways. The simplest (and most common) version, illustrated in Fig. 3.5, is **single-point crossover** in which a single crossover point is randomly chosen among the $m - 1$ possible points in a chromosome of length m. The procedure can be generalized to k-point crossover in which k crossover points are selected randomly and the chromosome parts are chosen from either parent[10] with equal probability. In **uniform crossover**, the number of crossover points is equal to $m - 1$.

These crossover methods are applicable both to binary and real-valued chromosomes. An additional crossover method applicable to real-valued chromosomes is **averaging crossover** in which a gene g taking the values g_1 and g_2 in the two parents takes the values

$$g_1 \leftarrow \alpha g_1 + (1 - \alpha)g_2 \quad (3.21)$$

and

$$g_2 \leftarrow (1 - \alpha)g_1 + \alpha g_2 \quad (3.22)$$

in the offspring. α is a number in the range [0, 1]. All crossover procedures introduced thus far are instances of **length-preserving crossover**, which is applicable in cases such as function optimization when the structure of the system (e.g. the number of variables) being optimized is known and fixed. However, in many situations, this is not the case. An example that will be considered in Section 3.3 is linear genetic programming in which the crossover procedure is allowed to modify the size of the chromosomes.

3.2.1.4 Mutation

Mutations play the important role of providing new material for evolution to work with. While mutations rarely have a positive effect on fitness when they occur, they may bring advantages in the long run. In GAs, the value of the mutation probability p_{mut} is usually set by the user before the start of the GA run, and is thereafter left unchanged. Typical values for p_{mut} are c/m, where c is a constant of order 1 and m is the chromosome length, implying that on average around one mutation occurs per chromosome. A brief motivation for this choice can be found in Appendix B,

---

[10] In obvious analogy with biology, the selected individuals in an EA are sometimes referred to as the **parents**, and the new individuals that are generated are referred to as the **offspring**.

Section B.2.5. However, note that mutations are normally carried out on a gene-by-gene basis, i.e. the decision whether or not to mutate is considered independently for each gene. Thus, in principle, the number of mutated genes can range from 0 to m.

Varying mutation rates can, of course, also be applied. However, if a varying mutation rate is used, one immediately encounters the problem of specifying exactly *how* it should vary. Thus, constant mutation rates are more commonly used. In case of binary encoding schemes, mutation normally consists of bit-flipping, i.e. a mutated 0 becomes a 1, and vice versa. By contrast, in real-number encodings, the modifications obtained by randomly selecting a new value in the allowed range (e.g. [0, 1]) are typically much larger than in the binary case, because each real-valued gene encodes more information than a binary-valued one. Thus, an alternative approach, known as **real-number creep**, is frequently used instead. In real-number creep, the mutated value (allele) of a gene is centered on the previous value, and the **creep rate** determines how far the mutation may take the new value, as illustrated in Fig. 3.10. Mathematically, in creep mutation, the value g of the gene changes according to

$$g' \leftarrow \psi(g), \tag{3.23}$$

where ψ is a suitable distribution, for instance a uniform distribution as in the left panel of Fig. 3.10, or a normal distribution as in the right panel of the same figure. In the case of a uniform distribution, the mutated value can thus be obtained as

$$g \leftarrow g - \frac{C_r}{2} + C_r r, \tag{3.24}$$

where r is a uniform random number in [0, 1] and C_r (the creep rate) is the width of the distribution.

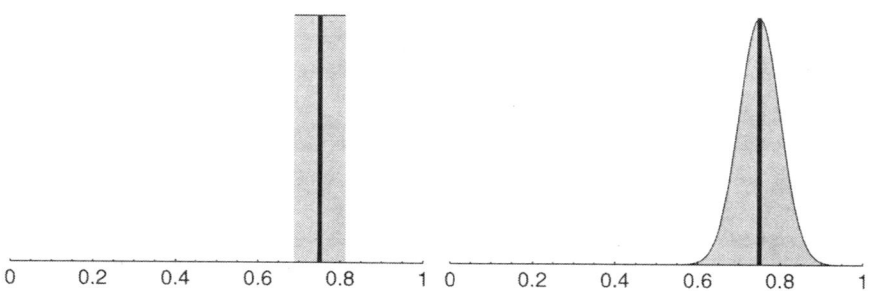

Figure 3.10: *Creep mutation using a uniform distribution (left panel) or a normal distribution (right panel). The initial value (equal to 0.75) of the gene g, before mutation, is illustrated as a thick black line. The mutated value is selected from the distributions shown in grey. For both distributions, the area of the shaded region is equal to 1. Random numbers following a normal distribution can be obtained using the Box–Müller transform described in Appendix C.*

In case a value outside the allowed range (e.g. [0, 1]) is obtained, which may happen if, say, g is close to one of the limits (0 or 1) and the distribution $\psi(g)$ is sufficiently wide, the mutated value is instead set to the limiting value.

3.2.1.5 Replacement
In Algorithm 3.1, generational replacement was used, meaning that all individuals in the evaluated generation were replaced by an equal number of offspring generated by applying selection, crossover and mutation. While common in GA applications, generational replacement is not very realistic from a biological point of view. In nature, different generations coexist, and individuals are born and die continuously, not only at specific intervals of time. By contrast, in generational replacement, there is no direct competition between individuals from different generations. Note, however, that there may still be some *de facto* overlap between generations, if the crossover and mutation rates are sufficiently low, because it is likely that some of the offspring will then be identical to the parents. However, this is a situation that one obviously wishes to avoid because re-evaluation of identical individuals does not improve the results obtained by the GA.

In general, replacement methods can be characterized by their **generational gap**, which simply measures the fraction of the population that is replaced in each selection cycle. Thus, for generational replacement, $G = 1$. In **steady-state replacement**, G equals $2/N$, i.e. two individuals are replaced in each selection cycle. In order to keep the population size constant, NG individuals must be deleted. In steady-state replacement, the two new individuals are typically evaluated immediately after being generated, and the two worst individuals, either among the original N individuals or among the $N + 2$ individuals (including the offspring), are deleted.

3.2.1.6 Elitism
Even though the best individual in a given generation is very likely to be selected for reproduction, there is no guarantee that it will be selected. Furthermore, even if it is selected, it is probable that it will be destroyed during crossover. In order to make sure that the best individual is not lost, it is common to make one or a few exact copies of this individual and place them directly, without any modification, in the new generation being formed, a procedure known as **elitism**. If the population size N is even, and if a single copy of the best individual is retained as the first individual in the new population, $N - 2$ new individuals are formed via the usual sequence of selection, crossover and mutation. The final individual, completing the new population, is formed using asexual reproduction, i.e. only selection and mutation.

3.2.1.7 A standard genetic algorithm
By summarizing the discussion above one can define a standard GA, see Algorithm 3.2. Note that the standard GA is quite similar to the one introduced in Algorithm 3.1; the only difference is the introduction of a crossover probability p_c and elitism. The choice of selection operator is not specified, meaning that both roulette-wheel selection and tournament selection are appropriate selection

1. Initialize the population by randomly generating N binary strings (chromosomes) c_i, $i = 1, \ldots, N$ of length $m = kn$, where k denotes the number of bits per variable.
2. Evaluate the individuals:

    2.1. Decode chromosome c_i to form the corresponding variables x_{ij}, $j = 1, \ldots, n$.
    2.2. Evaluate the objective function f using the variable values obtained in the previous step, and assign a fitness value $F_i = f(\mathbf{x}_i)$.
    2.3. Repeat steps 2.1–2.2 until the entire population has been evaluated.
    2.4. Set $i_{\text{best}} \leftarrow 1, F_{\text{best}} \leftarrow F_1$. Then loop through all individuals; if $F_i > F_{\text{best}}$, then $F_{\text{best}} \leftarrow F_i$ and $i_{\text{best}} \leftarrow i$.

3. Form the next generation:

    3.1. Make an exact copy of the best chromosome $c_{i_{\text{best}}}$.
    3.2. Select two individuals i_1 and i_2 from the evaluated population, using a suitable selection operator.
    3.3. Generate two offspring chromosomes by crossing, with probability p_c, the two chromosomes c_{i_1} and c_{i_2} of the two parents. With probability $1 - p_c$, copy the parent chromosomes without modification.
    3.4. Mutate the two offspring chromosomes.
    3.5. Repeat steps 3.2–3.4 until $N - 1$ additional individuals have been generated. Then replace the N old individuals by the N newly generated individuals.

4. Return to step 2, unless the termination criterion has been reached.

Algorithm 3.2: *A standard genetic algorithm applied to the case of function maximization. See the main text for a detailed description.*

operators in the standard GA. Furthermore, the standard GA uses binary encoding, but it can easily be modified to operate with real-number encoding, in which case creep mutations should also be included in step 3.4. Note that the fitness of the best individual F_{best} in Algorithm 3.2 refers to the global best fitness value, i.e. the best value found so far, in any generation. This is so since a copy of the best individual found so far is made in each generation, see step 3.1.

Figure 3.11 shows the results of a typical GA run in which elitism was used. As can be seen in the figure, the maximum fitness values (F_{best}) rise rapidly at first, before entering a phase in which long periods of stasis are interrupted by sudden jumps.

3.2.1.8 Parameter selection

As is evident from the discussion above, applying a GA to an optimization problem involves a non-trivial selection of components and parameters, and even though

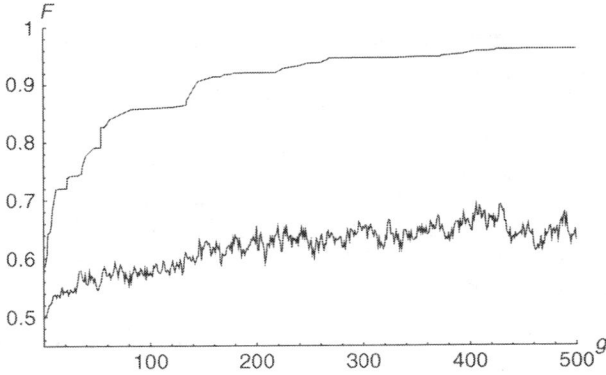

Figure 3.11: *A typical GA run for the maximization of the function $\Psi_5^{[10]}$ from Appendix D. The curves show the best (upper curve) and average fitness values obtained, as functions of the generation g. The fitness measure for this maximization task was simply taken as the function value. The population consisted of 100 individuals.*

there is no setup that is consistently superior over a wide range of problems, one can nevertheless exclude, for example, some particularly poor parameter choices. A few examples of such investigations will now be given.

Example 3.5
In this example, we shall compare roulette-wheel selection and tournament selection, keeping other operators and parameters fixed. A standard GA was implemented according to Algorithm 3.2 and was then applied to the problem of minimizing two of the benchmark functions introduced in Appendix D. Thirty bits per variable were used in the binary encoding scheme, giving a total chromosome length of $m = 30n$, where n is the number of variables, and m the chromosome length. The crossover rate was set to $p_c = 1.0$, and the mutation rate was set to $3/m$. For the two benchmark functions considered in this example, the (raw) fitness measures used were $F = 1/\Psi_1(x_1, x_2)$ and $F = 1/(1 + \Psi_2^{[3]}(x_1, x_2, x_3))$, respectively. Several different selection methods were tried. For each method, 3,000 runs were carried out on each benchmark problem, evaluating 10,000 individuals per run. For each run, the best fitness value obtained was stored.

Even though the distribution of the best result (from individual runs) is unknown, and probably quite different from a normal distribution, the distribution of the mean of the best result (over many runs) will approach a normal distribution, as described in Appendix C, Section C.1. The runs were therefore divided into 100 groups of 30 runs each. For each group, the average of the best result over the 30 runs was formed, giving one sample from this approximate normal distribution. The average and standard deviation of the 100 samples is given in Table 3.1. The estimates of the standard deviation s were obtained using eqn (C9).

From the table, it can be concluded that fitness ranking has a strong positive influence on the results obtained using roulette-wheel selection. For tournament selection, no particular conclusions can be drawn: a tournament size of five gives (slightly) better results than a tournament size of two for Ψ_1, but the opposite holds for Ψ_2. ∎

58 BIOLOGICALLY INSPIRED OPTIMIZATION METHODS

Table 3.1: *Comparison of selection methods, see Example 3.5. The first column lists the selection methods, and the remaining columns show the averages and estimated standard deviations as described in the example. Note that the listed standard deviations measure the variation (over 100 samples) in the mean (over 30 runs) of the best result obtained.*

Selection method	$\Psi_1 (x_1, x_2)$		$\Psi_2^{[3]} (x_1, x_2, x_3)$	
	Avg.	S.D.	Avg.	S.D.
RWS, no ranking	3.024	9.1×10^{-3}	1.2108	0.1811
RWS, with ranking	3.000	1.5×10^{-5}	0.3735	0.1069
TS, size $= 2$, $p_{\text{tour}} = 0.70$	3.003	1.4×10^{-3}	0.4253	0.1048
TS, size $= 2$, $p_{\text{tour}} = 0.90$	3.000	1.1×10^{-4}	0.3139	0.0870
TS, size $= 5$, $p_{\text{tour}} = 0.70$	3.000	0	0.6356	0.1822
TS, size $= 5$, $p_{\text{tour}} = 0.90$	3.000	0	0.7554	0.1872

RWS = *roulette-wheel selection*, TS = *tournament selection*.

Table 3.2: *Comparison of crossover probabilities. The first column lists the crossover probabilities, and the remaining columns show the averages and estimated standard deviations as described in Example 3.5.*

p_c	$\Psi_1 (x_1, x_2)$		$\Psi_2^{[3]} (x_1, x_2, x_3)$	
	Avg.	S.D.	Avg.	S.D.
0.00	3.00002	1.3×10^{-5}	1.020	0.183
0.25	3.00004	2.8×10^{-5}	0.655	0.139
0.50	3.00007	4.7×10^{-5}	0.516	0.130
0.80	3.00012	6.1×10^{-5}	0.394	0.118
0.90	3.00016	7.7×10^{-5}	0.389	0.118
1.00	3.00019	9.3×10^{-5}	0.363	0.108

Example 3.6
Using the same two benchmark functions as in Example 3.5, the effects of using different crossover probabilities were investigated for the case of single-point crossover. The population size, the number of bits per variable and the mutation rate were the same as in Example 3.5. Tournament selection was used, with a tournament size of two, and $p_{\text{tour}} = 0.90$. The raw fitness measures were the same as in Example 3.5. Three thousand runs, each evaluating 10,000 individuals over 100 generations were carried out. As in Example 3.5, the runs were divided into 100 groups of 30 runs each, giving 100 samples from the distribution of the mean. The results are summarized in Table 3.2. In this case, no particular conclusions can be drawn. For Ψ_1, the crossover operator appears to have a negative effect, whereas for Ψ_2, the best results are obtained using large values of p_c. ■

Table 3.3: *Comparison of mutation probabilities, see Example 3.7. The first column lists the mutation probabilities, and the remaining columns show the averages and estimated standard deviations as described in Example 3.5.*

p_{mut}	$\Psi_1(x_1,x_2)$		$\Psi_2^{[3]}(x_1,x_2,x_3)$	
	Avg.	S.D.	Avg.	S.D.
0	4.48910	0.75150	2.22610	0.505339
$1/2m$	3.03806	0.17743	1.35071	0.268959
$1/m$	3.00000	0.00000	1.18274	0.204092
$3/m$	3.00069	0.00041	1.23678	0.192616
$10/m$	3.04104	0.00984	1.70607	0.229125
1	4.81652	0.72717	1.27124	0.274613

Example 3.7
Again using the same two benchmark functions as in Example 3.5, the effects of using different (but constant) mutation probabilities were studied. The population size and the number of bits per variable were the same as in Example 3.5. The mutation rate was set as c/m, where m is the chromosome length (60 for Ψ_1, and 90 for $\Psi_2^{[3]}$), for different values of c. Tournament selection was used, with a tournament size of two, and $p_{tour} = 0.80$. The crossover probability was set to 0.80. The raw fitness measures were the same as in Example 3.5. Three thousand runs were carried out, each evaluating 10,000 individuals over 100 generations. As in Example 3.5, the runs were divided into 100 groups of 30 runs each, giving 100 samples from the distribution of the mean. The results, shown in Table 3.3, support the simple estimate for the optimal mutation rate ($1/m$) derived in Appendix B, Section B.2.5. ∎

3.2.2 Properties of genetic algorithms

The examples above, as well as Fig. 3.11, illustrate the convergence of the population in a GA towards highly fit individuals. However, the numerical examples say nothing specific about *how* GAs work. In this section, we shall address this issue, by briefly considering a few analytical models of GAs, starting with the **Schema theorem**.

3.2.2.1 The schema theorem
The schema theorem, derived by Holland [32], is one of the earliest results regarding the performance of GAs. Even though its importance has sometimes been exaggerated in the literature, it may be instructive to study it briefly. Consider a GA with a binary encoding scheme. As the name implies, the schema theorem relies on schemata (the plural form of the word *schema*) which are defined as strings consisting of 0s, 1s and wild-card symbols, here denoted x, which represent an arbitrary allele (either 0 or 1). Thus, for example, the schema 100xx1 represents

the strings 100001, 100011, 100101 and 100111. The schemata represent hyperplanes in the search space $\{0, 1\}^m$. Now, different schemata have different survival value. Consider, for example, a simple case in which the chromosomes encode two variables x_1 and x_2, using three bits per variable, and where the task is to maximize the function $e^{x_1 x_2}$. Using the standard decoding procedure for binary strings described above, but without any rescaling, the string 101111, say, will result in $x_1 = 2^{-1} + 2^{-3} = 0.625$ and $x_2 = 2^{-2} + 2^{-3} = 0.375$. In this trivial example, the schema $S_1 = 1111\text{xx}$ is clearly associated with higher fitness values than $S_2 = 00000\text{x}$, regardless of the values assigned to the wild cards. The schema theorem illustrates how certain schemata spread in the population from generation to generation. Letting $\Gamma(S, g)$ denote the number of copies of a schema S in generation g, it is easy to show (see Appendix B, Section B.2.1) that

$$E(\Gamma(S, g+1)) \geq \frac{\overline{F}_S}{\overline{F}} \Gamma(S, g) \left(1 - p_c \frac{d(S)}{m-1}\right) (1 - p_{\text{mut}})^{o(S)}, \qquad (3.25)$$

where $E(\Gamma(S, g+1))$ denotes the expectation value (see Appendix C) of $\Gamma(S, g+1)$, and $\overline{F}_S$ and $\overline{F}$ are the average fitness values of the schema S and of the entire population, respectively. $d(S)$ is the **defining length** of S, i.e. the distance between the first and the last non-wild-card symbol in S (thus, e.g. the schema $S = \text{x}01\text{x}1\text{x}$ has defining length $5 - 2 = 3$), and $o(S)$ is the **order** of S, i.e. the number of non-wild-card positions in S. p_c and p_{mut} denote the crossover and mutation probabilities, respectively, and m is the chromosome length. Eqn (3.25) says that schemata with above-average fitness, low defining length and low order (so called **building blocks**) tend to spread in the population, whereas schemata with the opposite characteristics tend to be eliminated. Thus, for example, if the low defining length, low order schemata xx10xxxxxx and xxxxxx11xx happen to be associated with high fitness, the number of such schemata will increase as more and more generations are evaluated.

While the schema theorem does give an indication of how building blocks spread in a population, it does so only in the form of an inequality. Furthermore, it falls short of elucidating how the building blocks are combined in order to form chromosomes associated with high fitness. Thus, the theorem has been severely criticized, perhaps justifiably so, by many authors, see for example Refs. [73] and [61].

3.2.2.2 Exact models

As an alternative to the schema-based analysis presented above, Vose [73] and others have considered exact models of GAs that, unlike the schema theorem, take into account, for example, recombination of chromosomes. In order to make the computations tractable, a simple GA, called SGA, was considered with binary chromosomes and fitness-proportional (i.e. roulette-wheel) selection followed by single-point crossover in which only one of the offspring is kept (to simplify the analytical calculations), and ordinary (bitwise) mutation.

In principle, all possible populations consisting of N individuals with chromosomes of length m can be enumerated and represented in a matrix P that, typically, will be very large: one can show (see Appendix B, Section B.2.2) that the number $v(N, m)$ of populations equals

$$v(N, m) = \binom{N + 2^m - 1}{2^m - 1}. \tag{3.26}$$

The matrix P is built in such a way that each column represents a population, i.e. a distribution of N chromosomes taken from the 2^m possible ones. As an example, consider the case $m = 3$, for which the possible chromosomes are 000, 001, 010, 011, 100, 101, 110 and 111. If, say, $N = 4$, a population with two copies of 001, one copy of 010 and one copy of 110 can then be represented as the column

$$\pi = (0, 2, 1, 0, 0, 0, 1, 0)^\mathrm{T}, \tag{3.27}$$

and so on. The evolution (over time) of the population will be a **Markov process**, i.e. a process in which the next state depends only on the state immediately preceding it. In order to study the progress of the population, one must compute the transition matrix T that determines the probability, under selection, crossover and mutation, of the transitions $\pi_i \to \pi_j$, where both i and j run from 1 to $v(N, m)$. Given the population $P(g)$ after g generations, the transition to $P(g+1)$ can be obtained as

$$E(P(g+1)) = TP(g). \tag{3.28}$$

The matrices involved will normally be truly huge, particularly because T has $v(N, m) \times v(N, m)$ elements. Even in the simple case considered above, with $N = 4$ and $m = 3$, the matrices P and T will have $8 \times 330 = 2{,}640$ and $330 \times 330 = 108{,}900$ elements, respectively. In a more realistic case with, say, $N = 100$ and $m = 30$, P and T will have around 1.4×10^{754} and 1.7×10^{1490} elements, respectively! Therefore, in view of the very large matrices involved, calculations involving exact models of this kind quickly become intractable, even though some general results regarding the dynamics of the SGA have, in fact, been obtained, see Ref. [61] for details.

However, a further simplification can be introduced: by letting the population size N tend to infinity, the effects of finite sampling can be ignored. Thus, consider a GA with infinite population size and chromosomes of (finite) length m so that the set Ω of possible chromosomes contains 2^m elements. Introducing an enumeration $j = 1, \ldots, 2^m$ of the strings, the probability distribution of the chromosomes can be written $p = p(j)$.

Now, because of the infinite population size, the exact probability distribution in the next generation can be obtained by applying an operator $\mathcal{G}(p)$ defined as a composition of three operators

$$\mathcal{G}(p) = \mathcal{G}_m(p) \circ \mathcal{G}_c(p) \circ \mathcal{G}_s(p), \qquad (3.29)$$

handling selection ($\mathcal{G}_s$), crossover ($\mathcal{G}_c$) and mutation ($\mathcal{G}_m$), respectively. Under fitness-proportional selection, the operator $\mathcal{G}_s$ can easily be obtained as

$$\mathcal{G}_s(p)(j) = \frac{F(j)p(j)}{\sum_{j \in \Omega} F(j)p(j)} \equiv \frac{F(j)p(j)}{\overline{F}}, \qquad (3.30)$$

where $\overline{F}$ denotes the average fitness. Crossover involves two individuals and can thus formally be represented as

$$\mathcal{G}_c(p)(j) = \sum_{k,l \in \Omega} C(j,k,l) p(k) p(l). \qquad (3.31)$$

The crossover operator is a bit tricky to represent explicitly. First of all, one must take into account that not all chromosomes can be generated through crossover of two given chromosomes. For example, upon crossing the chromosomes 1000 and 0110, one can never obtain, say, 0001. Thus, an explicit representation of a crossover operator must contain a mechanism for excluding chromosomes that cannot be generated. Let c_i denote chromosome i using the enumeration introduced above. One can show (see Appendix B, Section B.2.3) that $C(j,k,l)$ takes a non-zero value only if $(c_k \oplus c_j) \otimes (c_l \oplus c_j) = 0_m$, where $\oplus$ is the bitwise XOR operator and $\otimes$ is the bitwise AND operator, and 0_m denotes a chromosome consisting of m 0s.

The probability of generating any of the chromosomes that can be obtained as a result of crossing two given strings depends on the nature of the implemented crossover operator. In the simple case of uniform crossover in which a single offspring is generated by selecting genes from either parent, with equal probability, it is easy to see that the number of chromosomes that can be generated equals $2^{d_H(k,l)}$, where $d_H(k,l)$ denotes the **Hamming distance**[11] between the two parent chromosomes k and l. Because, in this case, all possible offspring are generated with equal probability, $C(j,k,l)$ takes the form

$$C(j,k,l) = \begin{cases} \frac{1}{2^{d_H(k,l)}} & \text{if } (c_k \oplus c_j) \otimes (c_l \oplus c_j) = 0_m, \\ 0 & \text{otherwise.} \end{cases} \qquad (3.32)$$

The final part of $\mathcal{G}$, the mutation operator $\mathcal{G}_m$ can be written as

$$\mathcal{G}_m(p)(j) = \sum_{k \in \Omega} M(j,k) p(k). \qquad (3.33)$$

---

[11] The Hamming distance between two binary strings is simply equal to the number of positions in which the strings differ.

In order to compute the matrix $M(j,k)$ we note that, in principle, any chromosome can mutate to any other chromosome. The probability of mutating a given chromosome k to form chromosome j is given by

$$M(j,k) = p_{\text{mut}}^{d_H(j,k)}(1 - p_{\text{mut}})^{n-d_H(j,k)}, \tag{3.34}$$

where p_{mut} is the mutation rate. Given the initial distribution p_1 of probabilities, the fitness function $F(j)$ and the operator $\mathcal{G}$, the exact distribution p_q for any generation q can, in principle, be computed exactly, albeit numerically.

In order to obtain a tractable expression for the transition from a given probability distribution p_q to the distribution p_{q+1} in the next generation, one may, as an additional simplification, consider **functions of unitation**, i.e. fitness functions that only depend on the number of 1s in the chromosome, regardless of their positions within the string. Such fitness functions are one-dimensional, and therefore easier to treat analytically than an m-dimensional fitness function. An example is given by the **Onemax** function $F(j) = j$, where the fitness simply equals the number (j) of 1s in the chromosome. Thus, in the Onemax problem, all chromosomes containing j 1s are, for the purposes of selection, equal. Changing the notation a bit, we let $p_q(j)$ denote the distribution of chromosomes with j 1s in generation q. Note that from this point onward j runs from 0 to m, rather than from 1 to 2^m as for the general, m-dimensional case above. Assuming random initialization, with equal probability of assigning 0s and 1s to the m genes, it is easy to show (see Appendix B, Section B.2.3) that the initial distribution $p_1(j)$ takes the form

$$p_1(j) = 2^{-m}\binom{m}{j}. \tag{3.35}$$

With the one-dimensional Onemax fitness function $F(j) = j$, the selection step, introduced in eqn (3.30), takes the form

$$\mathcal{G}_s(p)(j) = \frac{jp(j)}{\sum_{j=0}^{m} jp(j)} \tag{3.36}$$

With $p = p_1(j)$ as in eqn (3.35), using elementary properties of binomial coefficients (see Appendix B, Section B.2.3) the sum in the denominator can be computed as

$$\overline{F}_1 \equiv \sum_{j=0}^{m} jp_1(j) = 2^{-m}\sum_{j=0}^{m} j\binom{m}{j} = \frac{m}{2}, \tag{3.37}$$

where $\overline{F}_1$ denotes the average fitness in the initial generation. Thus, if both crossover and mutation are neglected so that $\mathcal{G} = \mathcal{G}_s$, the probability distribution $p_2(j)$ for the second generation (after one selection step) is obtained as

$$p_2(j) \equiv \mathcal{G}_s(p_1)(j) = 2^{1-m}\frac{j}{m}\binom{m}{j}. \tag{3.38}$$

This expression shows that the probability density is increased for chromosomes with above-average fitness ($j > m/2$) and decreased for chromosomes with $j < m/2$. Thus, as expected, fitness-proportional selection tends to increase the number of chromosomes associated with high fitness. Eqn (3.36) provides an exact quantitative measure of this increase. Proceeding in the same way, the distribution $p_q(j)$ after q generations can be determined,[12] given $p_{q-1}(j)$.

Crossover, however, cannot be considered without taking into account the exact positions of 0s and 1s in the two chromosomes being crossed. For example, even though the Onemax fitness function only depends on the number of 1s, the result of crossing, say, 100011 and 010011 will not be the same as the result of crossing 110100 and 111000. For simplicity, crossover will therefore be ignored here. As for mutation, an analytical expression for the matrix $M(j, k)$ can be derived (see Appendix B, Section B.2.3). However, the resulting expression is in the form of a rather complex sum. But, as mentioned in Section 3.2.1, mutation rates are typically set so that, on average, around one gene per chromosome mutates. Thus, in order to find closed-form expressions, we shall consider a greatly simplified mutation operator, such that, with probability p_μ, exactly *one*, randomly selected, gene mutates.

With this simplified mutation scheme, the distribution $p_2(j)$ resulting from selection and mutation can be written as (for a derivation, see Appendix B, Section B.2.3)

$$p_2(j) \equiv \mathcal{G}(p_1)(j) = 2^{1-m} \left(\frac{j}{m} + p_\mu \frac{m - 2j}{m^2} \right) \binom{m}{j}. \quad (3.39)$$

As expected, in the absence of mutations ($p_\mu = 0$), the expression for $p_2(j)$ is reduced to the one found in eqn (3.38). It is interesting to note the effect of mutations: evidently, mutations have a negative immediate effect because the mutation-related term in eqn (3.39) is negative for $j > m/2$. However, at least in the first generation, the positive effects of selection easily outweigh the negative effects of mutation. Let us now consider a combined analytical and numerical example.

Example 3.8
The simplest possible GA, using selection only, is applied to the Onemax problem. In this example, we shall investigate how the probability distribution $p_q(j)$ changes during the first few generations, i.e. for $q = 1, 2, \ldots$. In particular, the average fitness will be determined. Assuming random initialization of the binary chromosomes, the initial distribution $p_1(j)$ takes the form shown in eqn (3.35). Applying the selection operator $\mathcal{G}_s$, the distribution $p_2(j)$ is then obtained as in eqn (3.38). In order to obtain the distribution $p_q(j)$, the operator defined in eqn (3.36) can then be applied iteratively. For example, $p_3(j)$ is obtained as

$$p_3(j) \equiv \mathcal{G}_s(p_2)(j) = \frac{j p_2(j)}{\sum_{j=0}^{m} j p_2(j)}. \quad (3.40)$$

---

[12] Alas, a closed-form expression for $p_q(j)$, obtained directly from $p_1(j)$, cannot be obtained.

The sum in the denominator gives

$$\sum_{j=0}^{m} j p_2(j) = \frac{2^{1-m}}{m} \sum_{j=0}^{m} j^2 \binom{m}{j} = \frac{m+1}{2}, \quad (3.41)$$

where, in the final step, eqn (B23) has been used. This sum is also equal to the average fitness $\overline{F}_2$ of the second generation, cf. eqn (3.37). Returning to $p_3(j)$, we now obtain

$$p_3(j) = 2^{2-m} \frac{j^2}{m(m+1)} \binom{m}{j}. \quad (3.42)$$

In order to continue to later generations, one must compute sums of the form

$$S_k = \sum_{j=0}^{m} j^k \binom{m}{j}, \quad (3.43)$$

for $k = 3, 4, \ldots$ following, for example the procedure outlined in Appendix B, Section B.2.3, taking higher-order derivatives of both sides in eqn (B20). For the average fitness in the third generation, one obtains (check!)

$$\overline{F}_3 = \frac{m(m+3)}{2(m+1)}, \quad (3.44)$$

and for the fourth generation,

$$p_4(j) = 2^{3-m} \frac{j^3}{m^2(m+3)} \binom{m}{j}, \quad (3.45)$$

and

$$\overline{F}_4 = \frac{(m+1)(m^2 + 5m - 2)}{2m(m+3)}. \quad (3.46)$$

Evidently, the procedure can be continued up to any generation q, even though the expressions for $p_q(j)$ become more and more messy as q increases. Therefore, for higher values of q, a numerical approach can be applied using a large (but obviously finite) population size, applying standard roulette-wheel selection and the Onemax fitness function. An example of the results obtained using a chromosome length of 100 and a population size of 100,000 is shown in Fig. 3.12. Table 3.4 summarizes the numerical results and also shows the analytical results for small values of q.

As the GA progresses from one generation to the next, not only the average fitness increases, but also the probability of finding individuals with maximum fitness ($= m$). In the first generation, this probability equals $p_1(m) = 2^{-m}$, in the second generation it rises to $p_2(m) = 2^{-m+1}$ and in the third generation it reaches $p_3(m) = 2^{-m+2} m/(m+1)$, etc. ∎

To conclude this discussion we note that, with calculations of the kind presented above, it is possible to gain a better understanding of how GAs work, albeit in rather simplified cases. So far, however, we have said nothing about the expected running times, i.e. the expected number of evaluated individuals needed in order to achieve

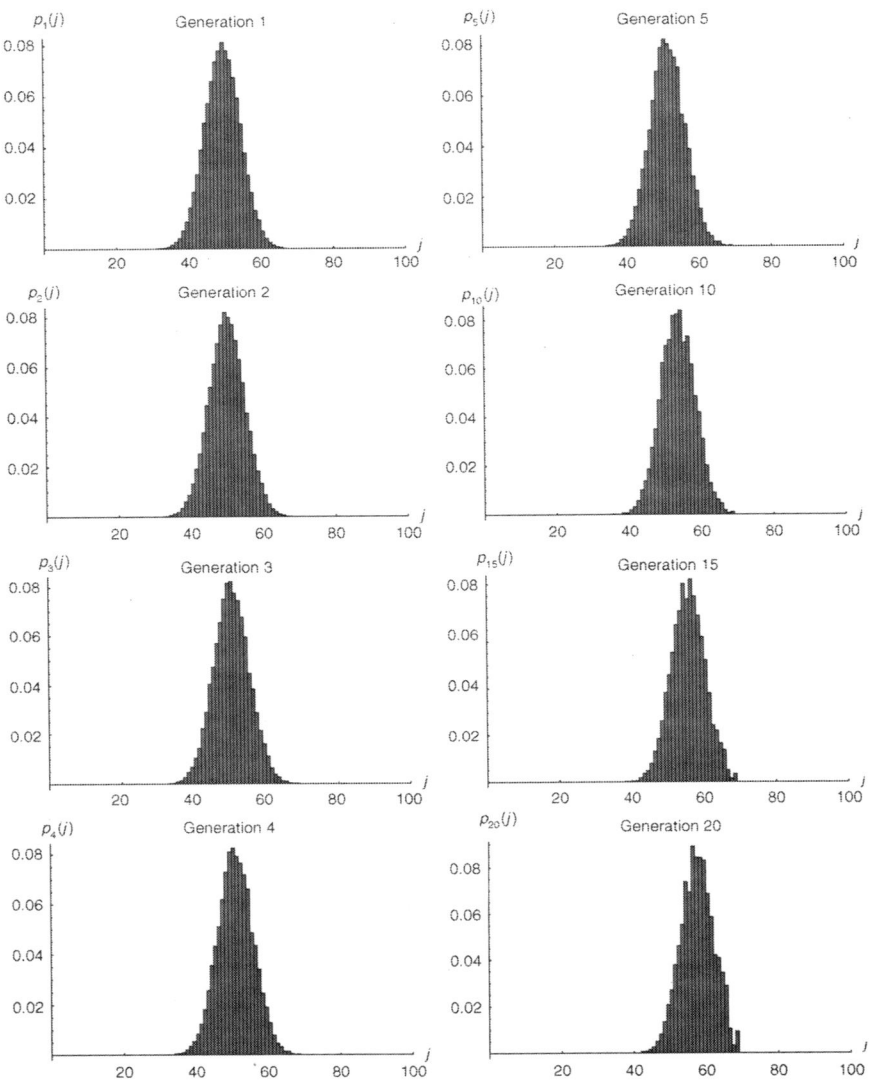

Figure 3.12: *Histograms over the probability distributions $p_q(j)$ for $q = 1, 2, 3, 4$ (left column, from top to bottom), 5, 10, 15, and 20 (right column), obtained using a simple GA with selection only, applied to the Onemax problem. The population size was 100,000 and the chromosome length was 100. The effects of the finite population begin to be noticeable around $q = 10$.*

Table 3.4: *Average fitness values for the Onemax problem, with $m = 100$. The columns marked $\overline{F}_q^{ex}$ show the exact average fitness values (computed up to $q = 5$), obtained as demonstrated in Example 3.8, and the columns marked $\overline{F}_q^{num}$ show the numerical results obtained with a population size of 100,000. Finally, the columns marked $p_q(m)$ show the probability of finding the best individual, with fitness m, computed up to $q = 5$.*

Generations	$\overline{F}_q^{ex}$	$\overline{F}_q^{num}$	$p_q(m)$	Generations	$\overline{F}_q^{ex}$	$\overline{F}_q^{num}$	$p_q(m)$
1	50.000	49.986	7.89×10^{-31}	5	51.943	51.916	1.19×10^{-29}
2	50.500	50.474	1.58×10^{-30}	10	–	54.096	–
3	50.990	50.983	3.12×10^{-30}	15	–	56.174	–
4	51.470	51.465	6.13×10^{-30}	20	–	58.032	–

a desired result. In order to investigate this issue, let us consider a very simple GA, with a population consisting of a single individual which is modified through mutations, and where a mutated individual is kept (thus replacing the parent) if the mutation is successful, meaning that the mutated individual obtains a higher fitness value than the parent. Assuming that the mutation rate is set according to

$$p_{\text{mut}} = k/m, \quad (3.47)$$

for some $k \ll m$, the expected number of evaluations $E(L)$, needed to obtain the maximum fitness ($= m$) for the Onemax problem, can be shown (see Appendix B, Section B.2.4) to be

$$E(L) \approx e^k \frac{m}{k} \ln \frac{m}{2}. \quad (3.48)$$

Example 3.9
In this example, we shall investigate the quality of the runtime estimate provided by eqn (3.48). The simple GA described in the derivation of this equation (see also Appendix B, Section B.2.4) was implemented in a computer program. For each combination of m and k, 10,000 runs were carried out, each run starting from a randomly generated chromosome. The runtime was estimated as the average number of evaluations over the 10,000 runs. Table 3.5 summarizes the results. As can be seen from the table, eqn (3.48) underestimates the running time slightly for small k, whereas for larger k, it overestimates the running time. Furthermore, as expected, one can note that the estimates improve as m becomes larger. ∎

3.2.2.3 Premature convergence

Typically, GAs are applied to problems for which the search space contains many local optima so that, consequently, the global optimum is hard to find. Because the initial population is normally generated randomly, most individuals are likely to have quite low fitness. However, as illustrated schematically in Fig. 3.13, while normally still far from the global optimum, some individuals are less bad than

Table 3.5: *Timing estimates for a simple GA compared with actual running times for different values of the chromosome length (m) and the mutation rate (k/m). E(L) is the estimated runtime (defined as the number of evaluations needed to obtain a chromosome with maximum fitness) according to eqn (3.48), whereas $\bar{L}$ is the average actual runtime over 10,000 runs. See Example 3.9.*

m	k	E(L)	$\bar{L}$	m	k	E(L)	$\bar{L}$
10	1	43.75	47.86	100	1	1063	1072
20	1	125.2	129.2	100	2	1445	1258
20	2	170.1	149.7	100	5	11612	7264
30	1	220.8	225.4	200	1	2504	2504
30	3	543.9	411.5	200	2	3403	2986
50	1	437.5	442.8	200	5	27338	17546
50	2	594.6	515.7	500	1	7504	7501
50	5	4777	3115	1,000	1	16893	16847

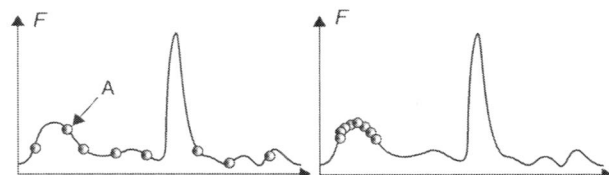

Figure 3.13: *A schematic illustration of premature convergence. The vertical axis measures the fitness F, and the horizontal axis is a simplified representation of the search space, which, of course, normally would be multi-dimensional. The individuals are shown as discs. In the left panel, one individual (A) is clearly better than the others, but still far from the global optimum. In the right panel, the population has converged on a suboptimal solution, close to the original location of individual A.*

others and, if the parameters and operators of the GA are not chosen wisely, such individuals can come to dominate the population. The individual marked A in the left panel of the figure happens to be situated fairly close to a local optimum in the search space. Because its fitness exceeds that of the other individuals, it will generate many offspring. Some generations will pass before it dominates the population, and during that time a fortuitous mutation may place some individual closer to another (and better) local optimum, or even the global optimum. However, if the optima are quite narrow, as in the figure, this is unlikely to occur. Instead, the entire population will gather around the local optimum. This convergence towards a suboptimal result is referred to as **premature convergence**, and it is one of the most common problems encountered in practical applications of GAs. The phenomenon can be compared with the loss of diversity associated with inbreeding in nature, a process that may occur in small populations and which generates less fit individuals, at least in the long run.

Once a population has gone through premature convergence, any subsequent change will be due solely to mutations; by itself, *crossover* between similar individuals produces similar offspring. In fact, in order to escape from the local optimum, an individual would have to undergo a so-called **macromutation**, consisting of a number of lucky changes occurring simultaneously. Clearly, such mutations are very unlikely, indicating that the search has more or less come to a halt after premature convergence has taken place (hence the name).

Fortunately, there exists several methods for avoiding premature convergence, some of which have already been introduced above. An example is the use of a crossover probability p_c smaller than 1. Because crossover is a very efficient operator, it should be used with caution in order to avoid the situation depicted in Fig. 3.13. Furthermore, if roulette-wheel selection is used, premature convergence can, to some extent, be avoided with the use of fitness ranking. In the left panel of Fig. 3.13, the ratio between the fitness of the best individual and that of the second best individual is, roughly, 2. In a population with N individuals, that ratio would be reduced to $N/(N-1) \approx 1$ (for normal population sizes of 30 individuals or more), using linear fitness ranking according to eqn (3.16). Note that, if tournament selection is used, fitness ranking is not needed.

Furthermore, the risk of premature convergence can also be decreased by introducing a mutation rate that *varies* with the diversity of the population defined, for example, as

$$D = \frac{2}{N(N-1)} \sum_{i=1}^{N-1} \sum_{j=i+1}^{N} \bar{d}(i, j), \qquad (3.49)$$

where N, as usual, denotes the population size, and where

$$\bar{d}(i, j) = \frac{1}{m} d(i, j) = \frac{1}{m} \sum_{k=1}^{m} \frac{|g(i, k) - g(j, k)|}{R_k}. \qquad (3.50)$$

Here, $g(i, k)$ denotes gene k in chromosome i and R_k is the range of gene k, introduced in order to handle cases in which different genes have different ranges. If binary encoding is used, $R_k = 1$ for all k and $d(i, j)$ is reduced to the Hamming distance $d_H(i, j)$. Clearly, after convergence (premature or otherwise), D will be small. Because mutations are random, they tend to increase the diversity in the population. Thus, in order to avoid premature convergence, the mutation rate can be increased if the diversity falls below a threshold $D_{\min}$. However, if the mutation rate is set too high, the search will be more or less random and the diversity is likely to stay above some threshold $D_{\max}$. Thus, the variation scheme for the mutation rate can be supplemented by a *decrease* in the mutation rate in such situations. Even though a varying mutation rate can reduce the risk of premature convergence, its use requires a detailed specification of exactly *how* the mutation rate should vary, as well as specific, numerical values for D_{min} and D_{max}. An example will be given below.

Premature convergence can also be mitigated by modifying the selection procedure, typically by means of some form of **mating restriction**. Such restrictions can be implemented in different ways, for example by preventing crossover between individuals that are deemed to be too similar. In this procedure, referred to as incest prevention, mating is allowed only between individuals for which $d(i, j)$, defined in eqn (3.50), exceeds a certain threshold d_{min}. Alternatively, one may use an **island model** in which the N individuals are divided into N_s subpopulations of $N_g = N/N_s$ individuals each. Clearly, without any exchange between the subpopulations, such an arrangement is equivalent to running N_s independent GAs with population size N_g. However, in island models, newly formed individuals are sometimes allowed to migrate to another subpopulation, for example, by every K^{th} generation removing the worst r% of the individuals in a randomly chosen subpopulation i and replacing them by the best r% of the individuals in another randomly chosen subpopulation j (with $j \neq i$).

Mating restrictions can also be implemented in the form of a **diffusion model** in which a topological arrangement of the individuals is introduced. An example is shown in Fig. 3.14. Mating restrictions are based on the placement of individuals on the grid shown in the figure. When an individual i, selected using one of the selection procedures described in Section 3.2.1, is to be replaced by a new individual, parents are selected only in the neighbourhood of the individual i. In

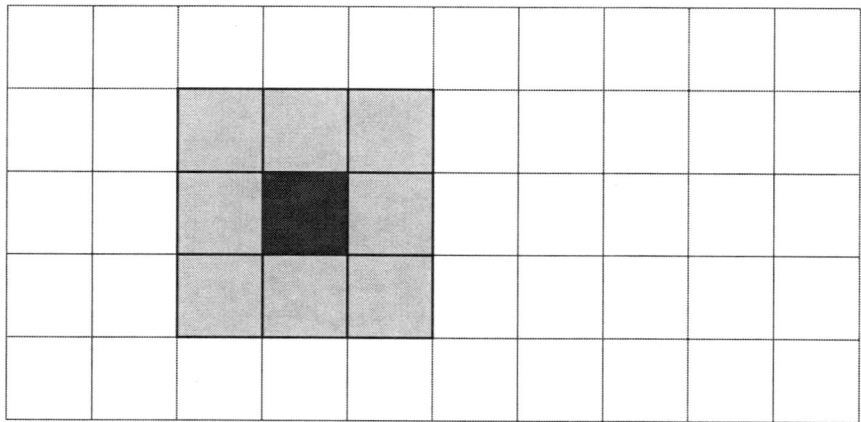

Figure 3.14: *Selection in a diffusion model GA with a population of 50 individuals arranged on a rectangular grid. A selected individual, shown as a dark square, is allowed to mate only with its immediate neighbours, shown in a lighter shade of grey.*

the figure, the neighbourhood contains eight individuals, but other neighbourhood sizes are, of course, possible as well. The topology of the grid can be chosen in many different ways. A common choice is to use **periodic boundary conditions** in which the two edges of a rectangular grid are joined to form a torus. To conclude this section, an example illustrating the prevention of premature convergence using varying mutation rates will be given.

Example 3.10
A GA is to be applied in order to minimize the function $\Psi_2^{[5]}$ (with five variables) defined in Appendix D. Investigate the effects on the population diversity and the performance of the GA when the mutation rate is varied based on the diversity measure defined in eqn (3.49).

Solution A standard GA was implemented according to Algorithm 3.2, using tournament selection, with a tournament size of two and with $p_{tour} = 0.80$. The chromosomes contained 40 bits per variables so that their total length ($= m$) equalled 200. The population size was set to 50, and a total of 20,000 individuals were evaluated per run (i.e. a total of 400 generations). The crossover probability was set to 1. Two sets of runs were carried out, each consisting of 500 runs. In the first set, the mutation rate was constant and equal to $0.005 (= 1/m)$. In the second set, a varying mutation rate was used based on the value of D which was measured in every generation before mutations were carried out. Specifically, p_{mut} varied according to

$$p_{mut} \leftarrow \begin{cases} p_{mut}\alpha & \text{if } D < D_{min} \\ p_{mut} & \text{if } D_{min} < D < D_{max} \\ p_{mut}/\alpha & \text{if } D > D_{max} \end{cases} \quad (3.51)$$

After some experimentation, the values $D_{min} = D_{max} = 0.25$ and $\alpha = 1.1$ were chosen. The left panel of Fig. 3.15 shows the variation in the diversity D from two typical runs, one from each set. Note that the initial diversity is around 0.5. For the runs in the first set, with constant mutation rate, D dropped from 0.5 to around 0.1. For the runs in the second set, D

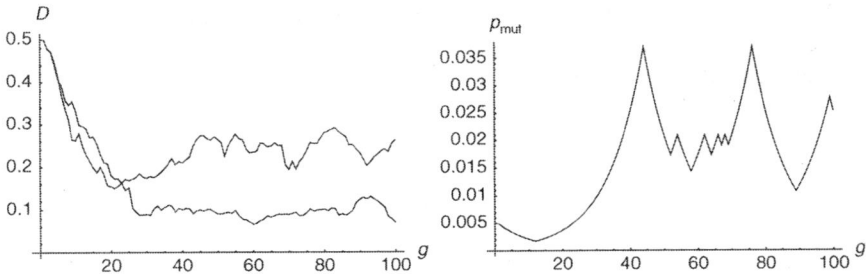

Figure 3.15: *Variation in diversity and mutation rates in two typical GA runs from Example 3.10. In the left panel, the lower curve shows the diversity (D) as a function of generation (g) for a run from the first set, whereas the upper curve shows D for a run from the second set, with vaying mutation rate according to eqn (3.51). The right panel shows the variation in mutation rate for the run from the second set.*

dropped to around 0.25, and then stayed around that value, as expected given the choice of $D_{\min}$ and $D_{\max}$. The right panel shows the variation of the mutation rate in a typical run from the second set (the same run for which the diversity was shown, in the left panel).

For the first set of runs, the average (over the 500 runs) of the best (= lowest) function value found was 31.04, whereas for the second set of runs, with varying mutation rate, the average equalled 12.13. The estimated standard deviations taken as the square root of the variance estimate obtained from eqn (C9) were 55.79 and 25.15, respectively, so that using Student's t-test (see Appendix C, Section C.1), with unequal variances, the null hypothesis $\mu_1 = \mu_2$ could be rejected at a p-value of around 0.002, where μ_1 and μ_2 represent the true averages of the distribution of results. Thus, in this case, the runs with variable mutation rate outperformed those with constant mutation rate, even though one should be careful before drawing far-reaching conclusions based on this example, particularly because the results (from the two sets of runs) most likely do not follow a normal distribution. ∎

3.3 Linear genetic programming

Even though GAs are perhaps the most commonly occurring type of EA used in applications, there are many other types as well, one of the most important being **genetic programming** (GP). In the original formulation of GP [39], tree-like structures representing, for example, a mathematical function were evolved. An example of such a structure is shown in Fig. 3.16. As can be seen in the figure, GP trees consist of **operators** taking one or several inputs, and **terminals** that do not take any inputs.

Many concepts from GAs, such as the presence of a population, the use of selection operators, etc., are the same in GP. However, some operators, such as crossover and mutation, must be modified for use in GP. Figure 3.17 shows an example of crossover, which, in GP, allows for a variation in the size of the evolving structures. Tree-based GP remains an active field of research, and has been used in many different applications, see Ref. [41]. An important application is function fitting, particularly in cases where the form of the function is unknown *a priori*. In such problems, where it is clearly not sufficient only to determine parameter values of a given function ansatz, the ability of GP to form arbitrary combinations of the available operators and terminals is crucial. For example, the GP tree in Fig. 3.16

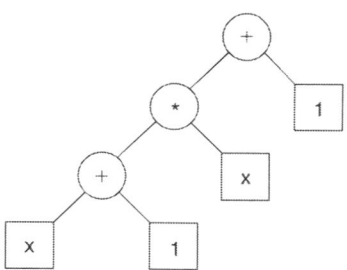

Figure 3.16: *A GP tree, which can be evaluated to give* $f(x) = (x+1)x + 1$.

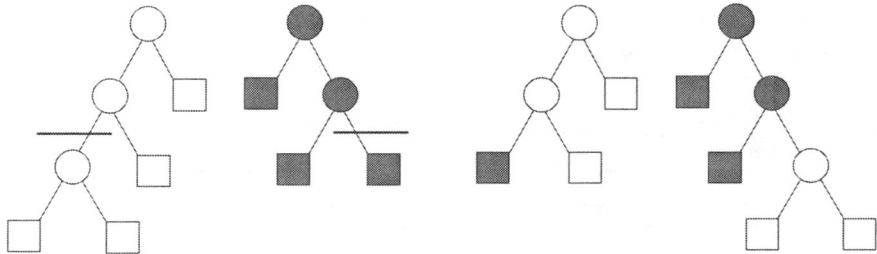

Figure 3.17: *Crossover in tree-based GP. Note that size variations are possible. The crossover points are indicated by thick horizontal lines.*

```
1    r_2 := r_1 + c_1
2    r_3 := r_2 + r_2
3    r_1 := r_2 × r_3
4    if (r_1 > c_2)
5        r_1 := r_1 + c_1
```

Figure 3.18: *A simple LGP program, with only five instructions. Variable registers are denoted r_i and constant registers are denoted c_i.*

can be evaluated to give $f(x) = x^2 + x + 1$. If the function being sought is, say, $f(x) = x^3 + x^2 + 1$, it can be obtained by a simple addition (which one?) to the tree, resulting from crossover with another tree.

Here, however, we shall consider a different version of GP, namely **linear genetic programming** (LGP) [6]. Unlike tree-like GP, LGP is used for evolving linear sequences (i.e. essentially computer programs) of basic instructions defined in the framework of a (low-level) programming language. A simple example of such a program is shown in Fig. 3.18.

3.3.1 Registers and instructions

Two central concepts in LGP are **registers** and **instructions**. Registers are of two basic kinds, **variable registers** and **constant registers**. Variable registers, denoted r_i in Fig. 3.18, can be used for providing input, manipulating data and storing the output resulting from a calculation. Thus, for example, the first instruction in Fig. 3.18 adds the contents of a variable register r_1 and a constant register c_1, and places the results in a second variable register r_2.

The set of all allowed instructions is called the **instruction set** and it may, of course, vary from case to case. However, no matter which instruction set is employed, the user must make sure that all operations generate valid results regardless of the inputs. For example, division by zero must be avoided. Therefore, in

Table 3.6: *Examples of typical LGP operators. Note that the operands can either be variable registers or constant registers.*

Instruction	Description	Instruction	Description
Addition	$r_i := r_j + r_k$	Sine	$r_i := \sin r_j$
Subtraction	$r_i := r_j - r_k$	Cosine	$r_j := \cos r_j$
Multiplication	$r_i := r_j \times r_k$	Square	$r_i := r_j^2$
Division	$r_i := r_j / r_k$	Square root	$r_i := \sqrt{r_j}$
Exponentiation	$r_i := e^{r_j}$	Conditional branch	if $r_i > r_j$
Logarithm	$r_i := \ln r_j$	Conditional branch	if $r_i \leq r_j$

practice, **protected definitions** are used. An example of a protected definition of the division operator is

$$r_i := \begin{cases} r_j / r_k & \text{if } r_k \neq 0, \\ c_{\max} & \text{otherwise,} \end{cases} \quad (3.52)$$

where $c_{\max}$ is a large pre-specified constant. Note that the instruction set can, of course, contain operators other than the basic arithmetic ones. Some examples of common LGP instructions are shown in Table 3.6. The branching instructions are commonly used in such a way that the *next* instruction is skipped *unless* the condition is satisfied. Thus, the instruction on line 5 in Fig. 3.18 will be executed only if $r_1 > c_2$. Note that it is possible to use a sequence of conditional branches in order to generate more complex conditions. Clearly, conditional branching can be augmented to allow, for example, jumps to a given location in the sequence of instructions. However, as soon as jumps are allowed, one faces the problems of making sure that the program terminates and avoids jumping to non-existent locations. In particular, the application of the crossover and mutation operators (see below) must then be followed by a screening of newly generated programs, to ascertain that they can be executed correctly. Jumping instructions will not be considered further here.

3.3.2 LGP chromosomes

In LGP, sequences of instructions of the kind just described are specified in a linear chromosome. The encoding scheme must be such that it identifies the **operands** (e.g. r_1 and c_1 in the instruction shown on the first line in Fig. 3.18), the **operator** (addition, in this case) and the **destination register** (r_2). An LGP instruction can thus be represented as a sequence of integers that identify the operator, the destination register and the operand(s). For example, if only the standard arithmetic operators (addition, subtraction, multiplication and division) are used, they can be identified by the numbers 1, 2, 3 and 4. An example of an LGP chromosome, encoding the short program shown in Fig. 3.18, is shown in Fig. 3.19.

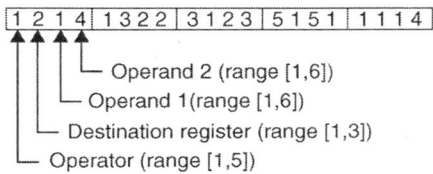

Figure 3.19: *An example of an LGP chromosome. In this example, it is assumed that there are three variable registers r_i, three constant registers c_i, and five operators o_i. See the main text for a complete description.*

Note that some instructions (e.g. addition) need four numbers for their specification, whereas others (e.g. exponentiation) need only three. However, in order to maintain a unified representation (to simplify the crossover procedure, see below), one may still represent each instruction by four numbers, simply ignoring the fourth number for those instructions that do not need it. The four numbers constituting an instruction may have different range. For example, the operands typically involve both variable registers and constant registers, whereas the destination register must be a variable register. In the specific example shown in Fig. 3.19, there are three variable registers available (r_1, r_2 and r_3) and three constant registers (c_1, c_2 and c_3), as well as five operators, namely addition, subtraction, multiplication, division and the conditional branching instruction 'if $(r_i > r_j)$'. Let $\mathcal{R}$ denote the set of variable registers, i.e. $\mathcal{R} = \{r_1, r_2, r_3\}$, and $\mathcal{C}$ the set of constant registers, i.e. $\mathcal{C} = \{c_1, c_2, c_3\}$. Let $\mathcal{A}$ denote the union of these two sets so that $\mathcal{A} = \{r_1, r_2, r_3, c_1, c_2, c_3\}$. The set of operators, finally, is denoted $\mathcal{O}$. Thus, in the set $\mathcal{O} = \{o_1, o_2, o_3, o_4, o_5\}$, the first operator ($o_1$) represents + (addition). An instruction is encoded using four numbers. The first number, in the range [1, 5], determines the operator as obtained from the set $\mathcal{O}$ and the second number determines the destination register, i.e. an element from the set $\mathcal{R}$. The third and fourth numbers determine the two operands taken from the set $\mathcal{A}$.

Before the chromosome can be evaluated, the registers must be initialized. In the particular case considered in Fig. 3.19, the constant registers were set as $c_1 = 1$, $c_2 = 3$ and $c_3 = 10$. These values then remain constant throughout the entire run, i.e. for all individuals. The variable registers should be initialized just before the evaluation of each individual. The input was provided through register r_1, and the other two variable registers (r_2 and r_3) were initialized to zero. The output could in principle be taken from any register(s). In this example, r_1 was used. The computation obtained from the chromosome shown in Fig. 3.19, in a case where r_1 (the input) was set to 1, is given in Fig. 3.20.

3.3.3 Evolutionary operators in LGP

With the exception of the crossover procedure and the range checking carried out during mutation (see below), LGP executes essentially in the same way as the

Genes	Instruction	Result
1, 2, 1, 4	$r_2 := r_1 + c_1$	$r_1 = 1, r_2 = 2, r_3 = 0$
1, 3, 2, 2	$r_3 := r_2 + r_2$	$r_1 = 1, r_2 = 2, r_3 = 4$
3, 1, 2, 3	$r_1 := r_2 \times r_3$	$r_1 = 8, r_2 = 2, r_3 = 4$
5, 1, 5, 1	if $(r_1 > c_2)$	$r_1 = 8, r_2 = 2, r_3 = 4$
1, 1, 1, 4	$r_1 := r_1 + c_1$	$r_1 = 9, r_2 = 2, r_3 = 4$

Figure 3.20: *Evaluation of the chromosome shown in Fig. 3.20 in a case where the register r_1 (the input register) was initially set to 1. The other variable registers, r_2 and r_3, were both set to zero, and the constant registers were set as $c_1 = 1, c_2 = 3$ and $c_3 = 10$. (Note that c_3 does not appear in this particular chromosome.) The first instruction (top line) is decoded from the first four genes in the chromosome. r_1 was designated as the output register, in this case giving an output $r_1 = 9$ at the end of the evaluation.*

standard GA described in Algorithm 3.2. In fact, once the instruction set and the registers (constant and variable) have been defined, the initialization of LGP proceeds in much the same way as in a GA: a population of chromosomes is generated by assigning random values, in the allowed range, to all genes keeping in mind that, in LGP, different genes may have different allowed ranges, as illustrated in Fig 3.19. Choosing the number of variable registers and the number of constant registers is non-trivial. Typically, only a few constant registers are defined because additional constants can always be built from those available in the constant registers. For example if, say, the product of two constants is needed, it can be generated by an instruction such as $r_1 := c_1 \times c_2$. The individuals, i.e. the computer programs, are decoded from the chromosomes and are then evaluated, a procedure that, naturally, differs from case to case (an example is given below). Selection then proceeds as in a GA. As for crossover, it is common to use a two-point crossover method that does *not* preserve the length of the chromosomes since there is no reason to believe that the length of the original, random chromosomes would be optimal. With two-point crossover, normally applied with crossover points between (rather than within) instructions, length variation is obtained. The crossover procedure is illustrated in Fig. 3.21. Mutations, finally, are carried out essentially as in a GA; the only difference is that the genes (in LGP) typically have different allowed ranges. To conclude the description of LGP, let us consider a specific example.

Example 3.11
An unknown function has been measured at $L = 101$ equidistant points in the range $[-2, 2]$. The resulting function values are shown in Fig. 3.22. Use LGP to find the measured function.

Solution An LGP program was implemented as described above. Both the number of variable registers and the number of constant registers were set to three, the constant registers taking the values $c_1 = 1, c_2 = 3, c_3 = -1$. Thus, the structure of the chromosomes was essentially equivalent to that shown in Fig. 3.19, except that in this example, the instruction set was taken as $\{+, -, \times, /\}$, encoded as the integers 1, 2, 3 and 4. During the evaluation

EVOLUTIONARY ALGORITHMS 77

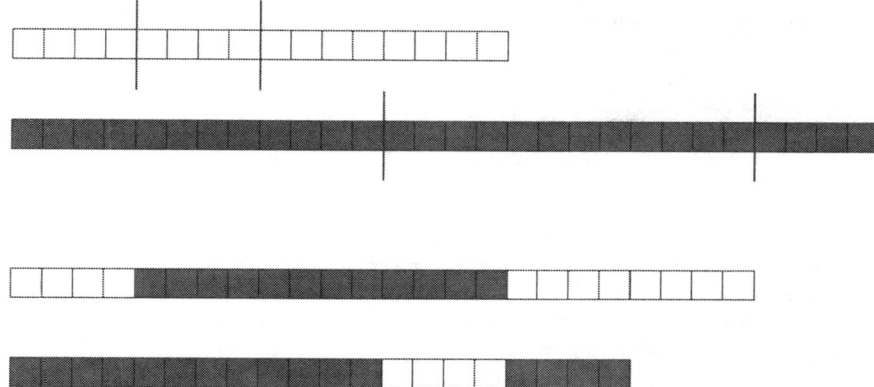

Figure 3.21: *An illustration of crossover in LGP. The two parent chromosomes are shown in the upper part of the figure and the two offspring are shown below. Note that two crossover points are chosen (in each chromosome). Crossover points are placed between instructions, each of which is represented by four squares.*

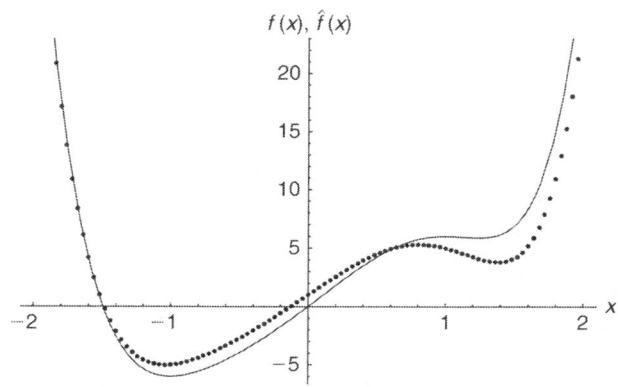

Figure 3.22: *The dots are data points measured from the function $f(x) = 1 + 8x - 3x^3 - 2x^4 + x^6$. The measured points were used in the function fitting example, using LGP, described in Example 3.11. A typical result found by the LGP program, namely $\hat{f}(x) = 9x + x^2 - 3x^3 + 2x^4 + x^6$, is shown as a solid line.*

of an individual, the 101 data points $f(x_i)$, $i = 1, \ldots, L$, where $x_i = -2 + 0.04(i-1)$, were considered one at a time, first setting the variable registers to $r_1 = x_i$, $r_2 = r_3 = 0$, and then running through the instructions defined by the chromosome and, finally, taking the output $\hat{f}(x_i)$ (for the data point in question) as the contents of r_1 at the end of the evaluation. Finally, when all L points had been evaluated, the **root-mean-square** (RMS) **error** was computed as

$$E_{\text{rms}} = \sqrt{\frac{1}{L}\sum_{i=1}^{L}(f(x_i) - \hat{f}(x_i))^2}. \qquad (3.53)$$

Finally, the fitness was formed as $F = 1/E_{\text{rms}}$. The standard evolutionary operators were used, namely tournament selection, with tournament size $= 5$ and $p_{\text{tour}} = 0.75$, and two-point crossover as illustrated in Fig. 3.21, with a crossover probability of 0.20. The mutation rate was 0.04. In the initial population, the number of instructions in the chromosomes ranged from 5 to 25. The population size was set to 100.

Of course, the example is a bit unrealistic in the sense that the data set was entirely noise-free, something that would never happen for an actual measurement of, say, a physical quantity. Nevertheless, as an exercise in genetic programming, it is interesting to see whether the correct function can be found. This function, unknown to the LGP program, was $f(x) = 1 + 8x - 3x^3 - 2x^4 + x^6$. Several runs were carried out, each lasting around 50,000 generations. In one case, the function $f(x)$ was found (exactly). Figure 3.22 shows a more typical result in which the best function found by the LGP program, with an error of $E_{\text{rms}} = 1.723$, was $\hat{f}(x) = 9x + x^2 - 3x^3 + 2x^4 + x^6$. It is shown as a solid line in the figure. Comparing $\hat{f}(x)$ and $f(x)$, one can see that, in this run, the LGP program found the leading term, but failed to fit exactly the lower-order terms. The function $\hat{f}(x)$ was obtained from a chromosome consisting of 13 instructions. ∎

3.4 Interactive evolutionary computation

In the algorithms described thus far, and indeed in those that will be introduced in later chapters, it has been assumed that a well-defined, quantitative fitness measure is available that can be computed objectively, irrespective of the preferences of the user. Thinking of optimization in a broad sense one can, however, picture situations where this might not be the case. One such example is design optimization of, say, a car. Obviously, there are objective criteria, such as limits on the weight, length, engine strength, etc. of the car that have to be taken into account, but there are also *subjective* criteria, such as visual appeal which is a crucial factor in determining the success of the car in the market. In such cases, one would need an algorithm in which the user is allowed to assign (or at least influence to some extent) fitness values based on subjective criteria. Algorithms of this kind have been developed in the EA subfield of **interactive evolutionary computation** (IEC).

In applications involving subjective criteria, rather than taking place in a mathematically defined fitness space, the search is confined to a kind of psychological space reflecting the user's preferences. Obviously, it is normally not possible to define a gradient in such a space, implying that gradient-based methods (as described in Chapter 2) cannot easily be generalized to cases involving subjective criteria. However, as EAs do not rely on gradients, they are well suited for this type of optimization.

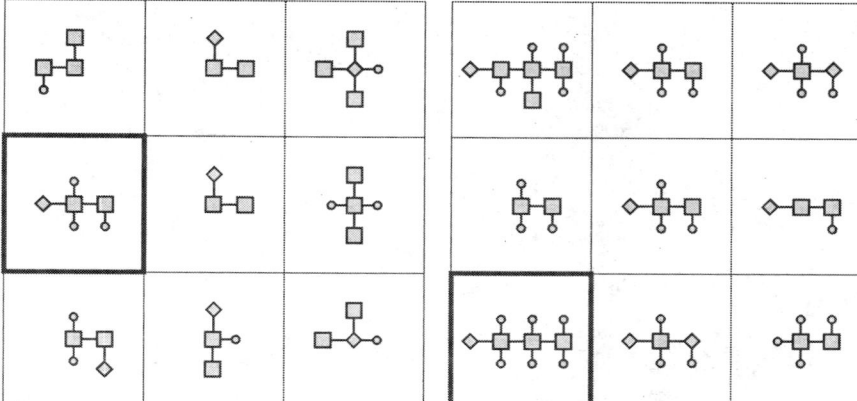

Figure 3.23: *A schematic illustration of IEC. Each of the nine images corresponds to a particular genotype. Based on subjective criteria (visual appeal), the user selects a preferred image, in this case the one that most resembles a caterpillar (the centre left image in the left panel), which is then placed at the centre of the screen in the next generation (right panel). Next, eight additional images are displayed. Those images are derived from chromosomes obtained by mutating the chromosome generating the preferred image. Of the nine images shown in the right panel, the lower left image is likely to be selected.*

In the simplest possible IEC algorithm, the user is presented with N images (individuals) on a computer screen (assuming, for now, that the optimization criterion is somehow based on visual appeal). The images are generated from chromosomes that encode, for example, the placement, length, number of copies, etc. of the different segments constituting the image. The user then selects the most appealing (or rather, in early generations, the least unappealing!) individual. The other $N-1$ individuals are discarded, the selected individual is placed at the centre of the screen, and is then surrounded by $N-1$ new individuals obtained through slight mutations of the selected individual, as illustrated in Fig. 3.23. The user then selects a new favourite and the process is repeated, each selection stage defining one generation. This method was used in the very first IEC algorithm, namely the **Biomorph** application introduced by Dawkins [11] as an illustration of the power of evolutionary change. Basically, Dawkins generated two-dimensional, black-and-white drawings, somewhat similar to the images appearing in Fig. 3.23, obtained through a recursive procedure (akin to the embryonic development of biological organisms), the shape of the resulting image depending on the values of nine genes. Even though the program was originally written to generate tree-like shapes, Dawkins soon discovered that a wide variety of forms (including images resembling spiders, foxes or even aircraft) could be generated through selective breeding over a surprisingly small number of generations.

In the time that has passed since Dawkins presented his Biomorphs, many examples of **evolutionary art**, essentially based on the same principles, but often

Figure 3.24: *An example of evolutionary art: "Mutation X Raycast". Artist: William Latham, 1993, www.williamlatham1.com. Copyright William Latham 1993. Reproduced with permission.*

including three-dimensional structures in colour [42, 63], have appeared. An example is shown in Fig. 3.24. Furthermore, the concept of **evolutionary music** has also emerged in which the user selects between music tunes rather than images. Also, more complex IEC algorithms have been developed in which the user may choose several individuals by assigning graded fitness values, normally quantized to a small set of allowed values (5–7 levels, say). Note that the user need not grade *all* individuals explicitly; individuals that are not graded are simply given a fitness value of zero. The chosen individuals are then used as parents when generating the next generation, through selection, crossover (commonly uniform crossover is used in IEC) and mutation. The typical flow of an IEC algorithm is shown in Algorithm 3.3.

The main problem in applying IEC algorithms is user fatigue, limiting the number of individuals that can be evaluated in a given generation. The limit is due both to display limitations (in the case of images), and the limits of human memory (e.g. in applications involving evolutionary music, where tunes are presented in sequence). Furthermore, user fatigue also limits the number of generations that can be evaluated: typically, IEC applications involve a maximum of 10–20 generations. On a positive note, though, one may observe that EAs often make rapid progress in early generations (see, e.g. Fig. 3.11).

1. Initialize the population by generating (from randomly initialized chromosomes) and displaying N images on a computer screen.
2. Subjectively evaluate the individuals and select the preferred image.
3. Generate the next generation:

    3.1. Make an exact copy of the preferred image and place it at the centre of the screen.

    3.2. Make $N - 1$ new individuals by mutating the selected individual. Display the $N - 1$ newly generated individuals, surrounding the individual placed at the centre of the screen.

4. Return to step 2, unless any of the displayed images are satisfactory.

Algorithm 3.3: *A basic IEC algorithm. See also Fig. 3.23.*

Several methods have been developed for reducing user fatigue in connection with IEC algorithms. First of all, the fitness assignment can be limited to a few levels, making the procedure more rapid and less tiring. For example, assigning either of the values 0 (bad), 1 (fair), 2 (good) or 3 (very good) is easier than assigning values on a scale involving, say, 100 different levels. Furthermore, it is possible to introduce a method for predicting the subjective fitness values assigned by the user so that larger population sizes can be used. The prediction can, for example, be based on the Euclidean distance in genotype space between a newly generated individual and one that has already been evaluated by the user. Thus, the user assigns non-zero fitness values to some (possibly all) of the N_u (10, say) individuals displayed. Next, N (100, say) new individuals are generated through selection, crossover and mutation, and their *estimated* fitness values are generated by the distance-based method just described. Finally, the N_u individuals with highest estimated fitness are displayed to the user, who again assigns fitness values, and the process is repeated. Another alternative for reducing user fatigue is to use a networked system that accepts inputs from several users. Of course, different users may assign different fitness values to a given individual but this might be favourable in, for example, design optimization: individuals that survive and spread their genetic material will be those that are favoured by many users. One should also note that in IEC applications it is rarely needed to reach a specific optimal point. Instead it is sufficient, or even desirable (e.g. in design optimization), just to reach an optimal *region* of the search space.

In addition to the evolutionary art and music described above, IEC has been applied to many different optimization problems [71] such as layout optimization for posters and web pages, industrial design (e.g. cars and other vehicles), fashion design, architectural design, face image generation (particularly for applications involving the generation of phantom images of criminal suspects) and robot control (e.g. gait generation for six- or eight-legged robots).

3.5 Biological vs. artificial evolution

Our description of biological evolution above was simplified, largely leaving aside such aspects as pairing of chromosomes, which allows concepts such as recessive traits, and the presence of **introns**[13] that leads to the appearance of alternative splicing of genes (i.e. different ways of assembling the various parts of an mRNA molecule). Nevertheless, as should be clear from the description of EAs above, in evolutionary optimization several additional simplifications are made, making such algorithms a mere caricature of their biological counterpart.

One such simplification involves the way chromosomes are used in evolutionary algorithms. The most common usage is illustrated in Fig. 3.4, where the chromosome is used as a lookup table from which the traits of the corresponding individual are obtained. In the simple case shown in the figure, the individual obtained consists of two variables x_1 and x_2 that can be used, for example, in a function optimization problem. The process taking place during the embryological development of a biological organism is vastly more complex.

First of all, in biological systems, the chromosome is *not* used as a simple lookup table. Instead, genes interact with each other to form complex **genetic regulatory networks** in which the **activity** (i.e. the level of production of mRNA) of a gene often is regulated by the activity of several other genes [59]. During embryological development of an individual, the genome thus executes what amounts to a complex program, resulting in a complete individual. An example of gene regulation is illustrated in a simplified fashion in Fig. 3.25. Here, a repressor protein, itself the product of a gene, may attach itself to an **operator** close to another gene, and thereby affect, i.e. increase or decrease, the ability of RNA polymerase to bind to the DNA molecule at the starting position of the gene (the **promoter region**). As mentioned in Section 3.1, genes that regulate other genes are called regulatory genes. Some regulatory genes regulate their own expression, forming a direct feedback loop, which can act to keep the activity of the gene within specific bounds. Gene regulation can occur in other ways as well. For example, a regulatory gene may be involved in the activation of a protein (i.e. the product of another gene) which then, in turn, may affect the transcription of other genes.

Furthermore, in evolutionary algorithms, the individual resulting from the decoding procedure is usually fixed during its evaluation. However, in biological systems, the genome remains active throughout the life time of the individuals, and continues to produce the proteins needed in the body at varying rates depending on the current needs. As mentioned in Appendix A, variation in gene regulation has been found to be crucial in (long-term) memory storage. Thus, even though evolution and learning certainly are different processes, the former occurring on vastly longer time scales than the latter, the processes are, to some extent, unified in their reliance on genetic information.

---

[13] Introns are non-coding regions that are removed from mRNA molecules before translation occurs. The parts that are used are referred to as **exons**.

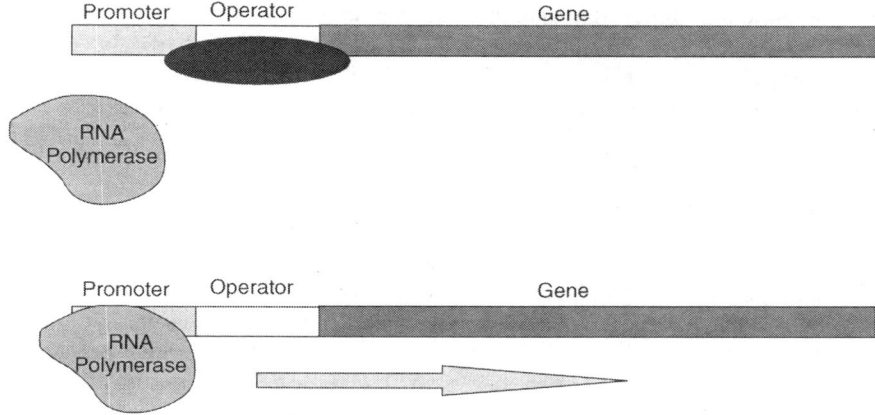

Figure 3.25: *A simplified representation of gene regulation. In the upper panel, a repressor protein, shown as a dark ellipse, is attached to the operator site, preventing RNA polymerase from binding to the promoter. In the lower panel, no repressor protein is present, and RNA polymerase can therefore attach itself and begin transcribing the gene.*

An additional simplification in most evolutionary algorithms is the absence of **multicellularity**. By contrast, in biological organisms, the development of an individual results in a structure with many cells (except, of course, in the case of single-celled organisms), and the level of gene expression in each cell is determined by its interaction with other cells. Thus, signalling between cells is an important factor in biological systems. Note, however, that the genome is the *same* in all cells in the body. It is only the *expression* of genes that varies between cells, determining, for example, whether a cell should become part of the brain (a neuron) or part of the muscle tissue.

3.6 Applications

In this section, a few applications of EAs will be described in some detail. Since EAs have been applied to hundreds or even thousands of different problems, any detailed description (of manageable length) of EA applications is bound to be incomplete. Admittedly, the applications considered here constitute a biased sample, reflecting the author's interests.

3.6.1 Optimization of truck braking systems

Downhill cruising of heavy-duty vehicles (such as large trucks) is a complex task, in view of the many different braking systems available on a modern truck: in addition to the standard pedal brakes (so called **foundation brakes**, abbreviated

FB), and the possibility of changing gears, the driver may also engage **auxiliary brakes** (ABs) such as the **engine brake** (EB) and the **compact hydrodynamical retarder** (CR) to reduce the speed of the vehicle. The distribution of the retardation force obtained from each source is of great importance because, for example, if the FBs are used too frequently, the brake discs may overheat, leading to a loss of braking power. Similarly, the cooling system for the ABs may become saturated, also resulting in a loss of braking power. Nevertheless, a standard strategy is to try primarily to use the ABs in order to avoid wear on the brake pads and discs. However, this strategy has been shown instead to cause excessive degradation of the drive tyres, resulting in high maintenance costs.

A driver may, of course, choose to enter a slope with a very low speed in order to avoid the problems listed above. However, this would lead to other problems, such as delayed delivery of goods and traffic congestion. Thus, ideally, the driver should try to maintain as high a speed as possible (taking safety into account) during down-hill cruising, while simultaneously avoiding overheating and excessive wear on drive tyres. This is indeed a non-trivial problem, especially because drivers may not know the current mass of the vehicle, and are normally not given continuous feedback regarding the temperature of the brake discs and the slope of the road. Thus, in countries with many steep hills, such as Switzerland or Norway, accidents caused by disc brake overheating have occurred.

In order to reduce the risk of such accidents, one may consider using a feed-forward neural network (FFNN, see Appendix A, Section A.2), taking as input the different variables that describe the state of the vehicle and its environment and giving as output the gear choice as well as the distribution of braking force between the FBs, EB and CR. Such an approach was employed in Ref. [45], where a GA was used for optimizing the neural networks. The choice of a GA rather than, say, backpropagation was motivated by the fact that, for this problem, suitable sets of input–output pairs (to guide the training of the FFNN) were not available; the FFNN must operate essentially in continuous time providing an output at all times,[14] and the result of applying a certain input may not be evident until many time steps later, thus making the definition of suitable outputs very difficult.

In Ref. [45], both the problem of maintaining a high mean speed without overheating the (disc) brakes and the problem of decreasing wear cost were considered. Here we shall content ourselves with considering the first case, i.e. increasing the mean speed as much as possible without risking an accident. The basic equation of (longitudinal) motion for the truck was taken as

$$M\dot{v} = F_{\text{drive}} + Mg \sin \alpha - F_{\text{res}} - F_{\text{FB}} - F_{\text{AB}}, \quad (3.54)$$

where M is the mass of the vehicle, v is its speed and α the current slope. F_{drive} is the propulsion force (assumed to be zero for downhill cruising), F_{res} is the resistance

---

[14] In practice, the differential equations describing the dynamics of the vehicle are of course discretized, using a time step dt, which should be smaller than the relevant time scale(s) of the problem. In this case, a time step of 0.01 s was used.

force combining air and roll resistance, and F_{FB} and F_{AB} are the FB and AB forces, respectively. The AB force can be further subdivided as

$$F_{AB} = F_{EB} + F_{CR}, \qquad (3.55)$$

where F_{EB} and F_{CR} denote the forces from the engine brake and compact retarder, respectively. The detailed equations describing F_{res}, F_{FB}, F_{EB} and F_{CR}, as well as the equations modelling the temperature variation in the brake discs and the coolant can be found in Ref. [45], and will not be given here. Suffice it to say that the problem is highly non-linear.

The neural networks were of fixed size, with five inputs and four outputs. The inputs consisted of (1) the vehicle speed, (2) the disc temperature, (3) the road slope, (4) the coolant temperature and (5) the engine speed. The outputs were taken as (1) the total retardation force request, (2) the force ratio between FBs and ABs, (3) the force ratio between EB and CR and (4) the gear choice. The standard logistic squashing function from eqn (A6) was used. Before being presented to the network, the input signals were rescaled to the interval [0, 1]. With the logistic squashing function, all output signals (denoted y_i^O, $i = 1, \ldots, 4$) from the network were also in the range [0, 1]. The first output signal (the retardation force request, F_{req}) was rescaled to an appropriate range. The second signal was used without rescaling, splitting the requested retardation force between the FBs and ABs. Thus, an output of $r \equiv y_2^O$ was interpreted such that a fraction r of F_{req} should be taken from the FBs, and a fraction $1 - r$ from the ABs. The third signal was interpreted in a similar way. The fourth and final output signal was used such that gears were shifted up if $y_4^O > 0.7$, shifted down if $y_4^O < 0.3$, and otherwise left unchanged. Note that the outputs from the FFNN represent *requested* actions that could not always be executed. For example, even though the FFNN may, in principle, request a gear change every time step, the frequency of gear shifts was, of course, limited roughly to one every other second.

A fairly standard GA was used, in which the chromosomes encoded the weights of the network, using real-number encoding. New individuals were formed using roulette-wheel selection, single-point crossover and mutations (both full-range mutations and creep mutations).

Each network, decoded from a chromosome, was tested by simulating the descent of the vehicle along a given road profile (see Fig. 3.26). The fitness measure was taken simply as the distance travelled by the vehicle during a maximum of τ_{max} s. The simulations were stopped prematurely, however, if any of the following constraints were violated: (1) $T_b < 500°C$, (2) $5 \text{ m/s} < v < 25 \text{ m/s}$, and (3) $600 \text{ rpm} < v_E < 2300 \text{ rpm}$, where T_b is the temperature of the brake discs, v the speed of the vehicle and v_E the engine speed. The simulation procedure is illustrated in Fig. 3.26.

First, some runs were carried out using roads with constant slope. It was found that the FFNNs, optimized by the GA, significantly improved performance: an increase in sustainable mean speed of 44%, compared to that achieved by a skilled human driver, was obtained on a 10% slope. On a 6% slope, the increase was 22%.

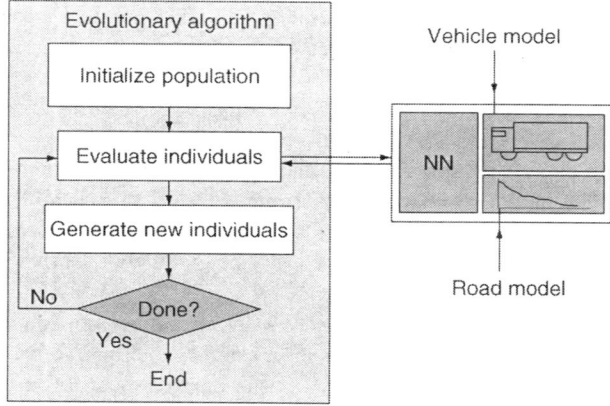

Figure 3.26: *Simulation procedure for the optimization of braking systems for heavy-duty trucks. The neural networks controlling the braking system were obtained from the chromosome, and the simulated truck was then allowed to descend a slope, following a given road model. Copyright: Vehicle System Dynamics (Taylor & Francis), accessible at www.informaworld.com. Reproduced with permission.*

Next, runs were carried out using road sections with varying slope, obtained by pasting together road data from the Kassel hills in Germany and from the French alps. Initially, rather short road sections were used but, as expected, it was found that the networks obtained in these cases were unable to cope with previously unseen road sections. Thus, this was a typical case of overfitting, see Appendix C, Section C.2. However, when trained against a very long road section (40 km), the resulting networks were able to generalize quite well, achieving only slightly worse results on test roads compared to the results obtained on the road used for training. An example of the variation of the road slope, vehicle speed, gear choice (note that the truck had 12 gears), AB retardation force and FB temperature over the first half of the training road is shown in Fig. 3.27. Note that the brake temperature approaches, but does not exceed, the cutoff (500°C) towards the end of this difficult, 20 km downhill road section.

To conclude, we can note that FFNNs optimized using GAs were able to improve driving performance, in many cases achieving better results even than experienced human drivers. Furthermore, unlike human drivers, who of course normally can see that a slope is coming before the descent begins, the FFNNs were not given any road profile preview, something that might have improved the results even more.

3.6.2 Determination of orbits of interacting galaxies

Collisions between galaxies constitute a spectacular astrophysical phenomenon. Often cataclysmic, these collisions distort the colliding galaxies, triggering the

EVOLUTIONARY ALGORITHMS 87

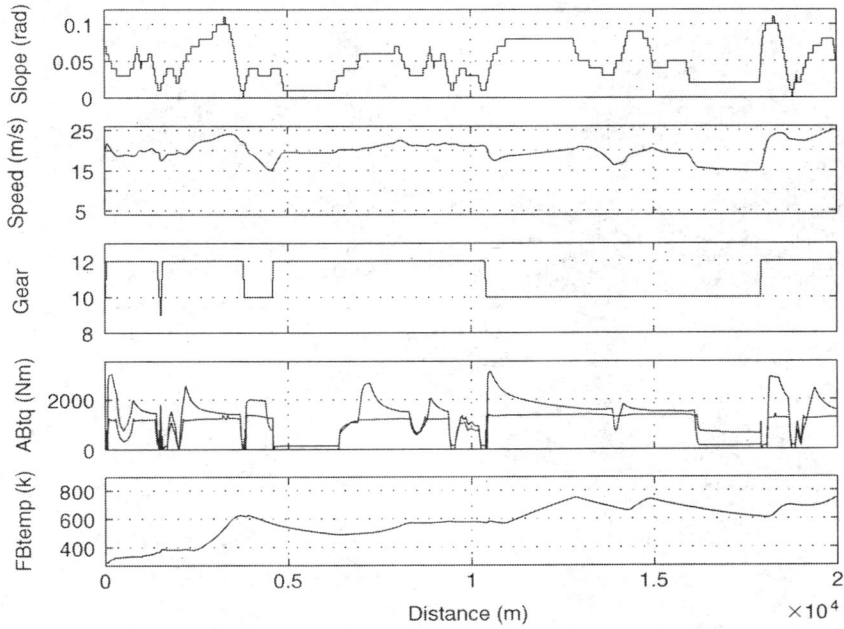

Figure 3.27: *An example of the results obtained with one of the best evolved FFNNs. From top to bottom, the first graph shows the slope as a function of horizontal distance, the second graph shows the speed, whereas the third and fourth graphs show the gear choice and AB torque (both obtained as outputs from the FFNN), respectively. The bottom graph measures the brake disc temperature. Copyright: Vehicle System Dynamics (Taylor & Francis), accessible at www.informaworld.com. Reproduced with permission.*

formation of tidal bridges and tails and also affect their spiral structure[15] (in the case of disc galaxies). A prime example of a galaxy collision is the NGC 5194/5195 system, also known as Messier 51 (M51), shown in Fig. 3.28, consisting of a large disc galaxy (NGC 5194) colliding with a smaller galaxy (NGC 5195). By human standards, galaxy collisions occur on very long time scales, typically several hundred million years. Thus, all that can be observed is essentially a snapshot, which hardly changes over a human life time. However, if the orbital parameters of the two colliding galaxies were known, one would be able to make computer simulations (so-called *n*-**body simulations** in which each galaxy is represented by a large number of point particles). A typical galaxy contains on the order of 10^{11}

---

[15] Galaxies are classified as either (1) disc galaxies that contain a flat disc, often with spiral arms, surrounded by an extended, ellipsoidal halo or (2) elliptical galaxies, which have an altogether ellipsoidal shape.

88 BIOLOGICALLY INSPIRED OPTIMIZATION METHODS

Figure 3.28: *The NGC5194/5195 (M51) system, consisting of a large spiral galaxy, seen almost face-on, and a smaller companion. Source: Telescopio Nazionale Galileo. Reproduced with permission.*

stars, too many to simulate even for modern computers. In an n-body simulation, where n typically is of order 10^5–10^6, each particle represents many stars.

As is well known from celestial mechanics, if the (relative) positions and velocities of a system consisting of two point particles are known, as well as the masses of the particles, their orbits can be fully determined. However, leaving aside the problem that galaxies are not point particles, observations of galaxies cannot provide all components of the relative position vector $\Delta \mathbf{x} = (\Delta x, \Delta y, \Delta z)^{\mathrm{T}}$ and the relative velocity vector $\Delta \mathbf{v} = (\Delta v_x, \Delta v_y, \Delta v_z)^{\mathrm{T}}$. In fact, only the position components in the plane of the sky (Δx and Δy) can be immediately observed as well as (using the Doppler effect) the velocity Δv_z in the direction perpendicular to the plane of the sky. This leaves three undetermined parameters, Δz, Δv_x and Δv_y. Furthermore, even though the masses of galaxies can be estimated (at least if the galaxy is not face-on, as seen from Earth) by studying the circular velocity of stars at different radii from the centre of the galaxy, the masses are rarely known with certainty. The problem of determining masses is also made more complicated by the (probable) presence of large quantities of dark matter, with a poorly known distribution. Finally, for severely distorted interacting (disc) galaxies, it might be difficult to determine the orientation of the discs, something that requires two angles, the inclination i and the position angle PA. Provided that the orbits of the two galaxies are integrated as if they were point particles, these two parameters will not affect the orbit. However, they *will* affect the detailed structure of the tidal bridges and tails formed during the collision. Thus, the number of unknown parameters ranges from three if masses and orientations are assumed to be known, up to nine if they are not.

Now, determining (without simulations) the likely structure of the tidal bridges and tails for a given set of orbital parameters is basically impossible. Thus, one is faced with a search problem of the kind mentioned in Section 2.5, requiring time-consuming simulations, essentially ruling out gradient-based search methods. In Ref. [74], a GA was instead used to search the space of relative galaxy orbits, using the following procedure. First, the unknown parameters determining the orbits were obtained from the chromosome, treating the galaxies as point particles. Next, once the orbits had been determined, combining the known values of Δx, Δy and Δv_z with the parameters obtained from the chromosome, two point particles were placed in the positions corresponding to the centres of the two galaxies. The point particles were then integrated *backwards* in time, until they were sufficiently separated not to have a strong tidal effect on each other. Next, a circular disc of particles was added to each galaxy (assuming that both were disc galaxies), and the two discs were integrated forward in time until the present time, using **restricted three-body simulations**, in which interparticle forces are neglected and the particles are thus only affected by the two point particles representing the total mass of each galaxy.[16]

Once the present time had been reached in the simulation, the mass distribution of the (now distorted) galaxies were compared with the actual mass distribution of the interacting system as obtained from observations,[17] and a fitness measure could thus be formed. In Ref. [76], the fitness measure was taken essentially as the inverse of δ defined as

$$\delta = \sum_{i,j} \left(m_{i,j}^{\text{obs}} + m_\epsilon \right) \left| \ln \left(\frac{m_{i,j} + m_\epsilon}{m_{i,j}^{\text{obs}} + m_\epsilon} \right) \right|, \quad (3.56)$$

---

[16] The alternative would be to include all interparticle forces, but such simulations would be much more time-consuming. Furthermore, interparticle forces are commonly of limited importance in violent interactions of the kind considered here. However, so-called **self-gravitating simulations** should preferably be carried out as a verification of the orbits found using a GA with restricted three-body simulations.

[17] Of course, observations of a galaxy yields a distribution of *light* rather than mass. However, it may be assumed that the so-called **mass-to-light ratio** is constant, allowing one to estimate the mass distribution.

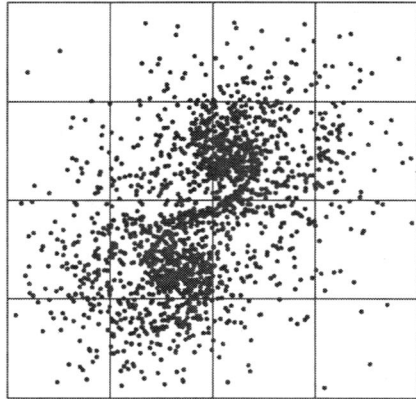

0.003	0.023	0.050	0.014
0.023	0.216	0.465	0.066
0.100	0.580	0.157	0.020
0.036	0.174	0.026	0.004

Figure 3.29: *An example of the mass distribution obtained from a simulation of two colliding galaxies, taken from Ref. [74]. The grid is superposed on the images, and the cells are assigned values corresponding to the mass in each cell (in this case, the total mass has been normalized to 2). For clarity, only a very coarse grid has been used in this figure. A similar grid can be superposed on a picture representing the actual, observed galaxies, and the deviation between the two distributions (δ) can then be obtained using eqn (3.56). Copyright: Astronomy and Astrophysics. Reproduced with permission.*

where $m_{i,j}$ and $m_{i,j}^{\text{obs}}$ denote the mass in cell (i,j) (see Fig. 3.29) obtained from the simulation and observations, respectively, and m_ϵ is a small, positive constant introduced to avoid problems for cells with $m_{i,j}^{\text{obs}} = 0$. Note that any subset of cells can be used in order to fit a certain part of the observed tidal structures while, for example, avoiding to model the central parts of the galaxies in detail. The fitness measure can also be augmented by a similar, grid-based measure of the deviation between the observed and simulated radial *velocities*, see Ref. [74] for details.

Using a standard GA, with roulette-wheel selection, single-point crossover and generational replacement, the procedure was applied to artificial data in Ref. [74]. A population size of 500 was used, and the runs lasted 100 generations in cases where the orientation was assumed to be known and 200 generations otherwise. It was found that the simulations more or less exactly reproduced the correct orbital parameters. In Ref. [76], the procedure was applied to the NGC5194/5195 system. This is one of the most studied galaxy pairs, and it had long been considered that the orbital plane of the system more or less coincides with the disc plane of the main galaxy, and that the beautiful spiral structure is a result of a recent passage by the smaller of the two galaxies (NGC 5195). However, in 1990, a giant arm of atomic hydrogen gas (HI, using astronomical terminology) was discovered [65], as shown in Fig. 3.30. The HI arm is truly enormous, and its presence rendered an orbit of the kind just described quite unlikely, since a recent passage would probably not have allowed the formation of such a giant arm. Thus, in Ref. [76] an unbiased search

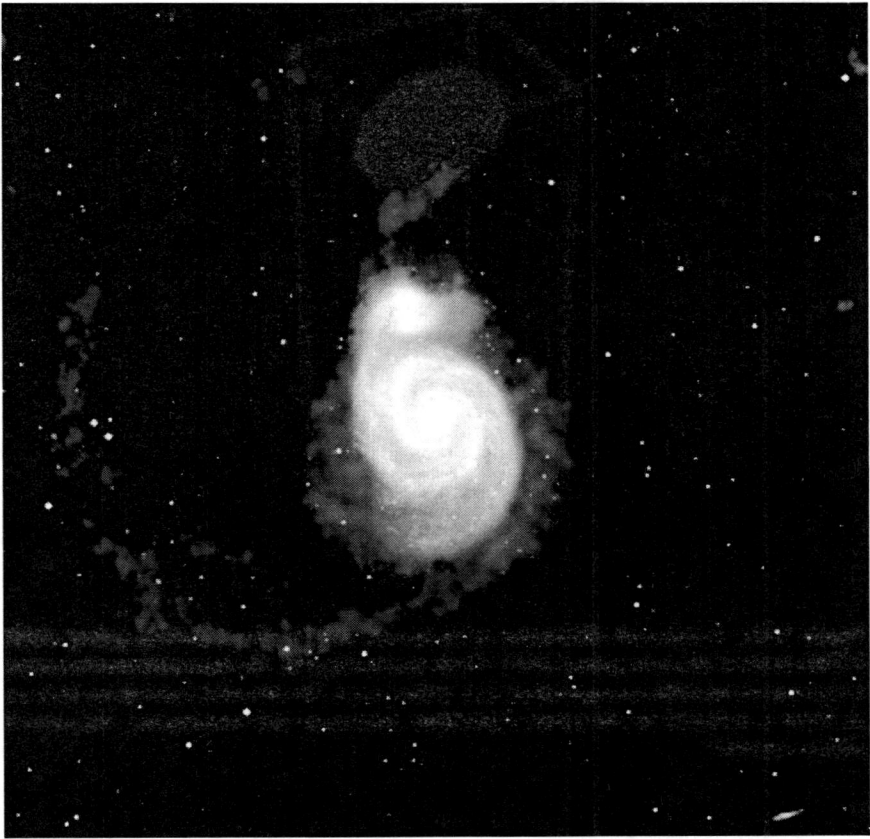

Figure 3.30: *The giant HI arm discovered by Rots et al. [65]. Note the difference in scale of this figure, compared to Fig. 3.28. Image courtesy of NRAO/AUI and Juan M. Uson, NRAO. Reproduced with permission.*

of the space of possible orbital parameters was carried out, using the GA described above. The results were startling: the best fit orbit, depicted in Fig. 3.31 turned out to be totally different from what had previously been assumed. According to the simulation results, the galaxy NGC 5195 is, in fact, located far behind (as seen from the Earth) NGC 5194, having plunged essentially perpendicularly through the disc around 850 million years ago. As mentioned above, the results were obtained using restricted three-body simulations. However, in order to strengthen the results, a fully self-gravitating simulation (including interparticle forces) was carried out, using the best-fit parameters obtained with the GA.

The results of this simulation, where 70,000 particles were used to represent NGC 5194, and where NGC 5195 was represented as a single point particle, are shown in Fig. 3.32. Comparing with Fig. 3.30, the large HI arm can be clearly

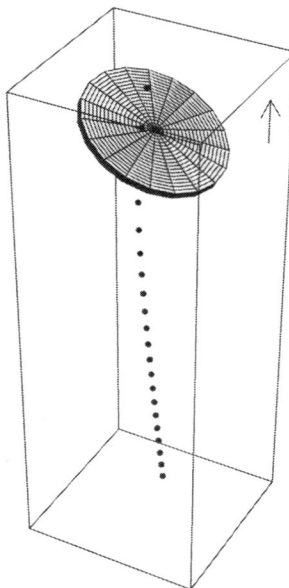

Figure 3.31: *A schematic illustration of the best-fit orbit for the perturbing galaxy (NGC 5195) relative to the main galaxy (NGC 5194) obtained in [76]. The interval between successive points on the orbit corresponds to 52.5 million years. The direction towards an observer on Earth is indicated by the arrow. Copyright: Astronomy and Astrophysics. Reproduced with permission.*

identified (no attempt was made, however, to reproduce the central parts of the two galaxies, visible in Fig. 3.28). Furthermore, the absence of an observed counter-arm, a common structure in galactic interactions, can be understood from the fact that even though such a structure is present in the simulations, it is quite diffuse and may thus be difficult to detect in observations.

Although a good fit was obtained for the mass distribution, the radial velocity field (not shown) obtained in the simulations deviated more severely from observations. Thus, while the study showed that the interaction had taken place in quite a different way than previously imagined, many questions concerning this fascinating celestial object still remain unanswered.

3.6.3 Prediction of cancer survival

During the last decade or so, the advent of **microarray** technology through which the expression levels (see Section 3.5) of thousands of genes can be measured simultaneously has dramatically increased the amount of data available to researchers in molecular biology. For example, in the case of diseases for which a genetic origin

EVOLUTIONARY ALGORITHMS 93

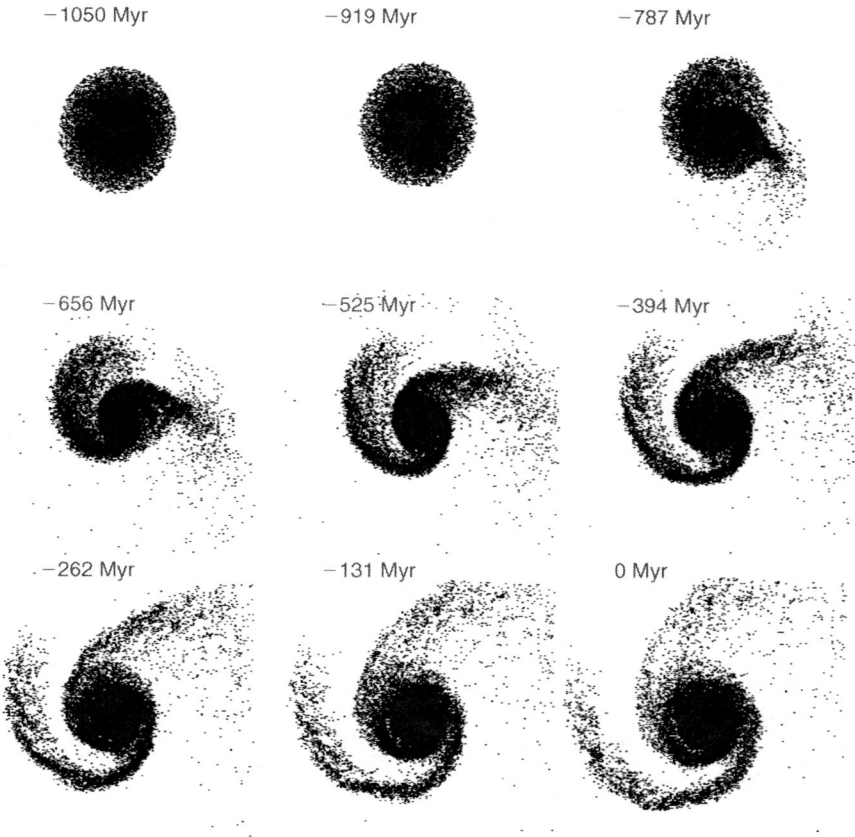

Figure 3.32: *A sequence of images from the self-gravitating simulation of the NGC5194/5195 system [76]. The images are equally spaced (131 million years) in time, which in this figure is counted backward from the present time. The perturbing galaxy passes through the plane of the main galaxy between the second and third frames. Copyright: Astronomy and Astrophysics. Reproduced with permission.*

is likely, the pattern of gene expression over a large part of the genome can provide crucial information regarding the nature and causes of the disease in question. As an example, the use of microarray-based gene expression measurements have been shown to improve the prediction of the clinical outcome, e.g. the survival time, in the case of breast cancer [72].

The data from such measurements is commonly presented in an $N \times M$ matrix, illustrated schematically in Fig. 3.33, where the rows enumerate the N features (the measured genes) $f_1, \ldots f_N$ and the columns represent samples $(S_1, \ldots, S_M)$ taken from M different patients. Thus, each sample S_i consists of a measurement

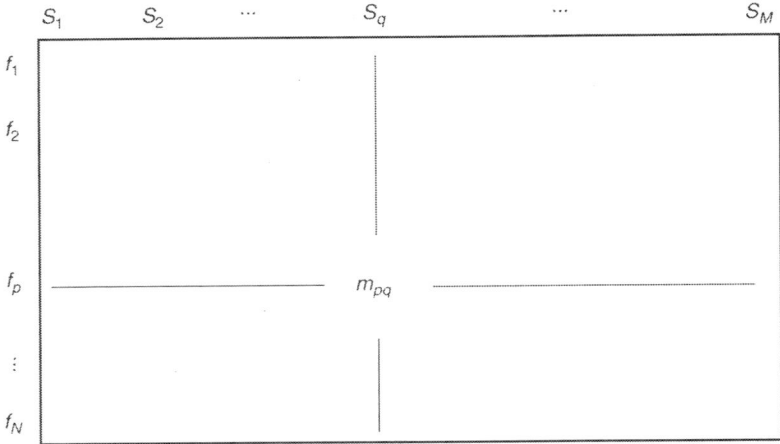

Figure 3.33: *A gene expression matrix, containing measurements of N genes (rows) for M samples (columns).*

of the expression level for N genes. The elements m_{pq} of the matrix are typically normalized to a given range, for example [0, 1].

Now, the data contained in a gene expression matrix can be used for different purposes. In **class discovery**, the goal may be, say, to discover new subclasses of a given disease, based on patterns of gene expression. Thus, in class discovery, samples with similar gene expression patterns are grouped together, commonly with the use of some form of **clustering algorithm**[18] [77].

By contrast, **data classification** is the procedure of finding those features (or genes, in the case of gene expression matrices) or combinations of features that best classify the samples. Thus, it is required that each sample should be assigned a class (i.e. a category). Therefore, for classification, the gene expression data matrix must be supplemented with a vector of class assignments. In general, the data may be divided into any number of classes. In the particular case of **binary classification**, there are two classes.

Mathematically, letting $i_1, \ldots, i_n$ denote the indices of the n features (among the N) used by a particular classifier, the goal is to find a function G that maps the pattern of gene expression $(f_{i_1}, \ldots, f_{i_n})$ for a given sample to the correct class C:

$$G(f_{i_1}, \ldots, f_{i_n}) \rightarrow C. \tag{3.57}$$

For example, a classifier using the two features f_3 and f_8 from the data matrix shown in Fig. 3.33 would have $i_1 = 3$ and $i_2 = 8$, and the classification rule would be

---

[18] Clustering is the procedure of dividing a data set into clusters (i.e. subsets) such that the members of each cluster share some relevant properties.

written as

$$G(f_{i_1}, f_{i_2}) \equiv G(f_3, f_8) \to C. \quad (3.58)$$

Thus, for example, for any sample q, the measured values m_{3q} and m_{8q} for features f_3 and f_8, respectively, would be used in the classifier. A classifier may use any subset of the N features available in the data matrix, and the number of features used may, of course, also be subject to optimization.

Many different approaches have been used to tackle classification problems of this kind, such as K-nearest-neighbour (KNN) classifiers and ANNs (see Appendix A). In the case of KNN, samples are assigned to a class C based on the class membership of its K nearest neighbours, obtained using the Euclidean distance measure in the space defined by the n selected features. Hybrid approaches exist as well: in Ref. [44], a GA was used for selecting n features, which were then used to carry out a KNN classification with $K = 3$.

Considering the case of predicting breast cancer survival times [72, 78], the goal is to map a given signature of gene expression to either of two classes, containing samples from patients with short and long survival time, respectively. An example of such a data set consisting of measurements of 5,277 genes for 97 patients, 51 of which belonged to the class (I) of long (over 5 years) metastasis-free survival and the remaining 46 samples to the class of short metastasis-free survival (class II), is given in Ref. [72], where also classifiers for this data set were generated. The method used in Ref. [72] started with the most relevant single feature (gene), i.e. the one that, by itself, provided better classification accuracy than any other feature, and added more and more features until no further improvement was detected. The classifiers obtained in Ref. [72] reached around 83% accuracy. However, the number (n) of features used was around 100. Now, in view of the great (but decreasing) costs involved in gene expression measurements, it would be preferable to generate a classifier using only a few (less than 10, say) features. If this were possible, diagnostic decisions based on the classifier could be confirmed by more traditional, independent methods that do not rely on microarrays.

Thus, in Ref. [78] a GA was employed in order to find linear classifiers, i.e. classifiers of the form

$$G = \beta + \sum_{j=1}^{n} \alpha_j f_{i_j}, \quad (3.59)$$

using a minimal set of features for the classification of the data set considered in Ref. [72]. Class assignments were such that samples for which $G > 0$ were assigned to class I, and samples with $G \leq 0$ were assigned to class II. Note that this is indeed a difficult search problem: the number $\nu(n, N)$ of ways of selecting n features among N equals

$$\nu(n, N) = \binom{N}{n}. \quad (3.60)$$

With $N = 5277$ and (say) $n = 5$, $v = 3.4 \times 10^{16}$ (!). In addition to selecting the n correct features, i.e. the integer indices $i_1, \ldots, i_n$, the GA also had to find appropriate values of the $n+1$ constants β and α_j, $j = 1, \ldots, n$. The chromosomes of the GA thus encoded a total of $2n+1$ parameters. Because the optimal number of features (n) was unknown *a priori*, the GA combined **parametric mutations**, changing the values (alleles) of the $2n+1$ parameters in a given chromosome with **structural mutations** (changing the *number* of features) that could either add or remove one feature in each mutation step. The parametric mutations were either full-range mutations or creep mutations. Furthermore, in order to speed up the search, the features were ranked based on their ability to classify the data set as single-feature classifiers ($n = 1$), and mutations were biased towards features with high classification accuracy. Several different fitness measures were used, the simplest one consisting of the fraction of correctly classified samples. In more complex fitness measures, an attempt was made to separate the two classes as much as possible, i.e. to place the n-dimensional hyperplane separating the two classes as far away as possible from the actual samples.

The results obtained with the GA showed that classifiers with as few as seven features could achieve 97.4% classification accuracy over training data (around 80% of the samples) and 89.5% over validation data (the remaining 20%), outperforming previously found classifiers. Other authors, for example [16] and [13], have also successfully employed GA-based methods in classification tasks involving gene expression matrices, and many other examples are sure to appear in the near future.

Exercises

3.1 Implement the basic GA described in Algorithm 3.1. Next, add elitism to the algorithm, and rerun the experiment described in Example 3.2. Compare the average best fitness value (after 25 generations) to the results obtained without using elitism, as in Example 3.2.

3.2 Consider a population consisting of five individuals with the fitness values (before ranking) $F_1 = 5$, $F_2 = 7$, $F_3 = 8$, $F_4 = 10$ and $F_5 = 15$. Compute the probability that individual 4 will be selected (in a single selection step) with (1) roulette-wheel selection, (2) tournament selection, with tournament size equal to 2, and the probability of selecting the best individual (in a given tournament) equal to 0.75, (3) roulette-wheel selection, based on linearly ranked fitness values, where the lowest fitness value is set to 1 and the highest fitness value set to 10.

3.3 Show by means of an example that the Gray code, introduced in Section 3.2.1, is *not* unique for $k = 4$.

3.4 Implement the standard GA described in Algorithm 3.2. Next, repeat the analysis in Example 3.5, including also Boltzmann selection, see eqn (3.18), the temperature T decreasing with the number of generations, and try to find a procedure for decreasing T, such that Boltzmann selection outperforms both roulette-wheel selection and tournament selection for the functions considered in the example.

3.5 Modify the standard GA algorithm so that it uses real-number encoding instead of binary encoding. Next, carry out two sets of benchmark runs, one with binary encoding and one using real-number encoding, to find the optimum of the sine-square function $\Psi_3(x_1, x_2)$ defined in Appendix D. What conclusions, if any, can be drawn regarding the choice of encoding scheme?

3.6 A GA is used for finding the maximum of the function $f(x_1, x_2) = x_1^2 + x_2^2$ in the interval $(x_1, x_2) \in [0, 0.9375]$. The fitness measure is taken simply as $f(x_1, x_2)$, without any rescaling or ranking. Assume that a binary encoding scheme is used, with four genes for the variable x_1 (the four first genes) and four genes for the variable x_2. Decoding is carried out in a simplified fashion where the first gene of the variable x_1 is multiplied by 2^{-1}, the second by 2^{-2}, etc., resulting in a value in the required range. The variable x_2 is obtained in a similar way. In the formation of new chromosomes, the crossover probability $p_c = 0.10$ is used, and the mutation rate is set to 0.01. In generation g, the population consists of six individuals with chromosomes 10101101, 01100111, 01110101, 01011001, 10010001 and 10001001. Use the schema theorem to estimate the number of copies of the schema $S_1 = 1xxx1xxx$ in generation $g + 1$.

3.7 Consider a chromosome with m binary-valued genes and assume that it is to be mutated using a mutation rate p. Determine the probability that the chromsome will undergo
  a) No mutation,
  b) Exactly one mutation,
  c) Less than three mutations.

3.8 A simplified GA, using only roulette-wheel selection, is applied to the One-max problem defined in Section 3.2.2, using random initialization of the (infinite) population consisting of binary strings of length m. Compute analytically the probability distribution in the third generation (after two selection steps) and show that the average fitness equals

$$\bar{F}_3 = \frac{m(m+3)}{2(1+m)}. \tag{3.61}$$

3.9 A GA similar to the one considered in the previous problem is used for finding the maximum of the function

$$F(j) = j(m-j), \tag{3.62}$$

where j is the number of 1s in the (binary) chromosomes and m is the chromosome length. Compute
  a) The average fitness in the first generation,
  b) The probability distribution $p_2(j)$ in the second generation, i.e. after evaluation and roulette-wheel selection,
  c) The average fitness in the second generation.

3.10 LGP is commonly used in function fitting problems where a given, unknown function has been sampled at a number of different points. If the fitted function is known to be a polynomial, with a degree no larger than d, one can also use a standard GA with fixed-length chromosomes, by setting the chromosome length such that it can represent all possible polynomials of degree d or less. However, the number of terms grows very rapidly with d. For a polynomial p of n variables and degree d, use the notation

$$p(x_1, x_2, \ldots, x_n) = a_{00\cdots 0} + a_{10\cdots 0}x_1 + a_{01\cdots 0}x_2 \\ + \cdots + a_{110\cdots 0}x_1 x_2 + \cdots + a_{00\cdots d}x_n^d. \quad (3.63)$$

As a specific example, a polynomial of two variables and degree $d = 2$ would be given as

$$p(x_1, x_2) = a_{00} + a_{10}x_1 + a_{01}x_2 + a_{20}x_1^2 + a_{11}x_1 x_2 + a_{02}x_2^2, \quad (3.64)$$

so that, in this case, there would be six parameters to determine. How many parameters would there be in a polynomial with
a) 3 variables and degree 2,
b) 4 variables and degree 3,
c) n variables and degree d.
Hint: for part (c), see the derivation of eqn (3.26) in Appendix B.

3.11 In the derivation in Appendix B, Section B.2.4, of the expected runtime for a simple GA applied to the Onemax problem, a simplified expression was used to estimate the probability of improving a given chromosome, i.e. increasing the number of 1s, see eqn (B38). As we shall see in this exercise, an exact expression can be derived.
a) Find an exact expression for the probability $P^+(m, j, k, p)$ of raising the fitness, in a single mutation step, from j to $j + k$ in a chromosome consisting of j 1s and $m - j$ 0s, assuming that the (bitwise) mutation rate is p and that $j < m$. Note that the number of 1s can increase in several different ways. For example, the fitness is raised from j to $j + 2$ if two 0s mutate to 1s, but also if, say, four 0s mutate to 1s and two 1s mutate to 0s. (The answer may contain a sum.)
b) Summing $P^+(m, j, k, p)$ over k, from 1 to $m - j$, one obtains the exact probability $P^e(m, j, p)$ of improving a chromosome with j 1s in a single mutation step. Compute $P^e(10, 1, 0.1)$, $P^e(10, 5, 0.1)$, and $P^e(10, 9, 0.1)$ and compare the results with those obtained using the approximate expression obtained from eqn (B38). What conclusions can be drawn?

Chapter 4

Ant colony optimization

Ants are among the most widespread living organisms on the planet, and they are native to all regions except Antarctica and certain islands (such as Greenland). Ants display a multitude of fascinating behaviours, and one of their most impressive achievements is their amazing ability to co-operate in order to achieve goals far beyond the reach of an individual ant. An example, shown in Fig. 4.1, is the dynamic bridge-building exhibited by Weaver ants. Ants of this species build nests by weaving leaves together. In many situations, leaves may be too far for one ant to put them together. In such cases, Weaver ants form a bridge using their own bodies. Another example is the collective transport of heavy objects, exhibited by many species of ants. In this case, groups of ants co-operate to move an object (i.e. a large insect) that is beyond the carrying capacity of a single ant. While perhaps seemingly simple, this is indeed a very complex behaviour. First of all, the ant that discovers the object must realize that it is too heavy to transport without the help of other ants. Next, in order to obtain the help of additional ants, the first ant must carry out a process of **recruitment**. It does so by secreting a substance from a poison gland, attracting nearby ants (up to a distance of around 2 m). Then, the transport must be co-ordinated regarding, for example, the number of participating ants, their relative positions around the object, etc. Furthermore, there must be a procedure for overcoming deadlocks, in cases where the transported object becomes stuck. Studies of ants have shown that their capability for collective transport is indeed remarkable. As an example, groups of up to 100 ants, carrying objects weighing 5,000 times the weight of a single ant have been spotted [50].

The complex cooperative behaviours of ants have inspired researchers and engineers. For example, the dynamic bridge-building just mentioned has inspired research on rescue robots. In this application, the scenario is as follows: freely moving robots are released in a disaster area to search for injured or unconscious victims of the disaster. If, during the search, a robot encounters an insurmountable obstacle,

100 BIOLOGICALLY INSPIRED OPTIMIZATION METHODS

Figure 4.1: *A group of Weaver ants bringing leaves together, using their own bodies as a dynamic bridge. Photo by Václav Skoupý. Reproduced with permission.*

it may recruit other robots to form a bridge, or some other suitable structure, to pass the obstacle, and then continue the search.

Ants are also capable of remarkably efficient foraging (food gathering). Watching ants engaging in this behaviour, one may believe that the ants are equipped with very competent leaders and are capable of communicating over long distances. However, neither of these assertions would be true: ants self-organize their foraging, without any leader, and they do so using mainly short-range communication. In fact, the essential features of ant foraging can be captured in a rather simple set of equations that underlie the **ant colony optimization** (ACO) algorithms that will be studied in this chapter. However, we shall begin with a brief discussion of the biological background of ACO.

4.1 Biological background

In general, co-operation requires communication. Even though direct communication (e.g. by sound) occurs in ants, their foremost means of communication is indirect. Ants are able to secrete substances known as **pheromones**, and thus to modify the environment in a way that can be perceived by other ants. This

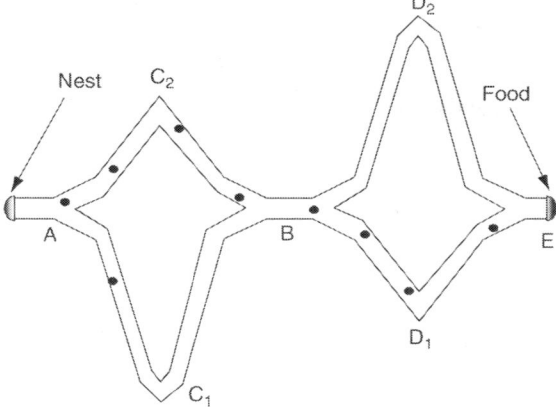

Figure 4.2: *A schematic view of the experimental setup used by Deneubourg et al. [15].*

form of indirect communication through modification of the local environment is known as **stigmergy** and it is exhibited not only by ants but by many other species (e.g. termites) as well.

During foraging, ants traverse very large distances, in some cases up to 100 m, before returning to the nest carrying food. This is an amazing feat, considering that the size of a typical ant is of the order of 1 cm. How is it achieved? The answer is that, while moving, an ant will deposit a trail of pheromone.[1] Other ants that happen to encounter the pheromone scent, are then likely to follow it, thus reinforcing the trail. The trail-making pheromones deposited by ants are volatile hydrocarbons. Hence, the trail must be continually reinforced in order to remain detectable. Therefore, when the supply of food (from a given source) drops, the trail will no longer be reinforced, and will eventually disappear altogether. The local communication mediated by pheromones is thus a crucial component in ant behaviour.[2]

The stigmergic communication in ants has been studied in controlled experiments by Deneubourg *et al.* [15]. In their experiments, which involved ants of the species *Linepithema humile* (Argentine ants), the ants' nest was connected to a food source by a bridge, shown schematically in Fig. 4.2. As can be seen from the figure,

---

[1] The rate of pheromone laying is not uniform, however. Typically, inbound ants (moving towards the nest) deposit more pheromone than outbound ones.

[2] In fact, pheromones have many other uses as well, both in ants and other species. A fascinating example is that of slave-making ants: about 50 of the 12,000 or so species of ants are slave-makers, i.e. they invade the nests of other ants and capture pupae containing workers of the invaded species. The exact invasion procedure varies between species, but some species are capable of using pheromones in a kind of chemical warfare: releasing their pheromones, they cause members of the other species to fight among themselves, allowing the invading species to go about its business.

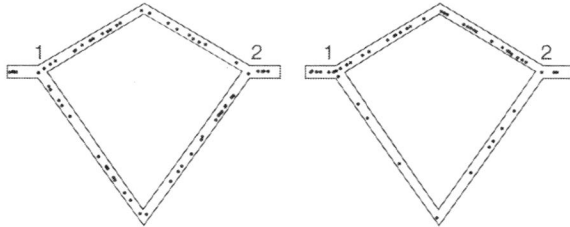

Figure 4.3: *A simplified setup, used in numerical experiments of ant navigation. The ants move from the nest at the left edge, following either of two possible paths to the food source at the right edge, before returning to the nest, again selecting either of two possible paths. The left panel shows a snapshot from the early stages of an experiment, where the probability of an ant choosing either path is roughly equal. Later on, however (right panel), the simulated ants show a clear preference for the shorter path.*

an ant leaving the nest is given a choice of two different paths, at two points (A and B) along the way towards the food source. The second half of the path might seem unnecessary, but it was included to remove any bias. Thus, at the first decision point, the short path segment is encountered if a left turn is made whereas, at the second decision point, an ant must make a right turn in order to find the short path segment. Furthermore, the shape of the branches (at the decision points A, B and E) was carefully chosen to avoid giving the ants any guidance based on the appearance of the path. Both branches involved an initial 30° turn, regardless of which branch was chosen.

Thus, the first ants released into this environment made a random choice of direction at the decision points: some ants would take a long path (A–C_1–B–D_2–E), some a short path (A–C_2–B–D_1–E) and some a path of intermediate length. However, those ants that happened to take the shortest path in both directions would, of course, return to the nest before the other ants (assuming roughly equal speed of movement). Thus, at this point, ants leaving the nest would detect a (weak) pheromone scent, in the direction towards C_2, upon reaching A, and would therefore display a preference for the shorter path. During their own trek, they would then deposit pheromone along this path, further reinforcing the trail. In the experiment, it was found that, a few minutes after the discovery of the food source, the shortest path was clearly preferred.

In addition to carrying out the experiment, Deneubourg *et al.* also introduced a simple numerical model for the dynamics of trail formation, which we will now consider briefly. Since the model deals with artificial ants, there is no need to use the elaborate path considered in the experiment with real ants. Thus, a simplified path with a single decision point in each direction, shown in Fig. 4.3, can be used. The numerical model accounts for the fact that *L. humile* deposit pheromones both on the outbound and the inbound part of the motion.

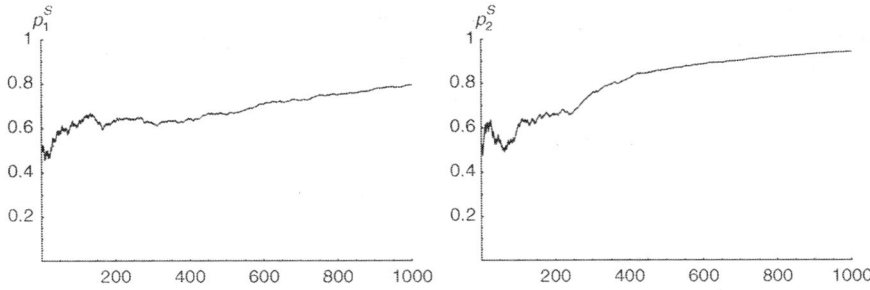

Figure 4.4: *The probabilities p_1^S (left panel) and p_2^S plotted as functions of the number of ants passing each decision point (1 and 2, respectively) for a typical run. The details will, of course, vary from run to run, since the ants select paths probabilistically.*

Thus, upon arriving at decision point 1, while moving towards the food source, an ant chooses the short path with a probability p_1^S, empirically modelled as

$$p_1^S = \frac{(C+S_1)^m}{(C+S_1)^m + (C+L_1)^m}, \quad (4.1)$$

where S_1 and L_1 denote the amount of pheromone along the short and long paths, respectively, at the decision point, and C and m are constants. Furthermore, before proceeding, the ant deposits a unit of pheromone on the chosen branch. Since the ants must choose one of the two paths, the probability p_1^L of selecting the long path equals $1 - p_1^S$. Similarly, while inbound (i.e. moving towards the nest), arriving at decision point 2 the ant chooses the short path with probability

$$p_2^S = \frac{(C+S_2)^m}{(C+S_2)^m + (C+L_2)^m}, \quad (4.2)$$

and then again deposits a unit of pheromone on the chosen branch.

In the experiments with real ants, traversing a short branch (e.g. A–C_2–B in Fig. 4.2) took around 20 s, whereas an ant choosing a long branch would need $20r$ s, where r is the length ratio of the two branches. It was found in Ref. [15] that a good fit to experimental data (for various values of r in the range [1, 2]) could be found for $C = 20$ and $m = 2$.

Example 4.1
The model for ant navigation described in eqns (4.1) and (4.2) was implemented in a computer program, and the experiments described in Ref. [15] were repeated. The results obtained, using the parameter values $C = 20$, $m = 2$ and $r = 1.5$, are shown in Figs. 4.3 and 4.4. The number of simulated ants was equal to 100. The left panel of Fig. 4.3 shows a typical distribution of ants (shown as black dots) in the early stages of a simulation. As can be seen in the figure, the distribution is rather uniform, reflecting the fact that, initially, the simulated ants have no preference when selecting paths. However, in the right panel, a snapshot taken

after a couple of hundred ants had traversed the paths, it can be seen clearly that the short path is preferred. In Fig. 4.4, the probabilities p_1^S and p_2^S are plotted, for a typical run, as functions of the number of ants having passed each point. ∎

4.2 Ant algorithms

Given the biological background of ACO, i.e. the foraging behaviour of ants, it is, perhaps, not surprising that ACO algorithms operate by searching for paths (representing the solution to the problem at hand) in a graph. Thus, applying ACO is commonly a bit more demanding than using a GA (or PSO, for that matter; see Chapter 5) since, in order to solve a problem using ACO, it is necessary to formulate it as the problem of finding the optimal (e.g. shortest) path in a given graph, known as a **construction graph**. Such a graph, here denoted as $\mathcal{G}$, can be considered as a pair $(\mathcal{N}, \mathcal{E})$, where $\mathcal{N}$ are the nodes (vertices) of the graph, and $\mathcal{E}$ are the edges, i.e. the connections between the nodes. In ACO, artificial ants are released onto the graph $\mathcal{G}$, and move according to probabilistic rules that will be described in detail below. As the ants generate their paths over $\mathcal{G}$, they deposit pheromone on the edges of the graph. Thus, each edge e_{ij}, connecting node j (denoted v_j) to node i is associated with a pheromone level τ_{ij}. Note that, in ACO, the construction graphs are, in general, directed, so that τ_{ij} may be different from τ_{ji}.

The exact nature of the construction graph $\mathcal{G}$ varies from problem to problem. The left panel of Fig. 4.5 shows a typical construction graph for one of the most common applications of ACO, namely the **travelling salesman problem** (TSP). The aim, in TSP, is to find the shortest path that visits each city once (and only once) and, in the final step, returns to the city of origin. The number of possible paths, which equals $(n-1)!/2$ where n is the number of nodes, grows very rapidly as the size of the problem is increased. Returning to the left panel of Fig. 4.5, in standard TSP, the nodes $v_i \in \mathcal{N}$, represent the cities, and the edges e_{ij} the (straight-line) paths between them. Thus, for this problem, the construction graph has a very straightforward interpretation: it is simply equivalent to the **physical graph** in which the actual movement takes place. In fact, because of its straightforward interpretation, the TSP is an ideal problem for describing ACO algorithms, and we will use it in the detailed description below.

However, in many problems, the construction graph is an abstract object, and functions merely as a vehicle for finding a solution to the problem at hand. The right panel of Fig. 4.5 shows a **chain construction graph**, which generates a binary string that can then, for example, be used as the input to a Boolean function. A chain construction graph that generates a binary string of length L consists of $3L+1$ nodes and $4L$ edges. Starting from the initial node v_1, either an up-move or a down-move can be made, the decision being based on the pheromone levels on the edges e_{21} and e_{31}. If an up-move is made (to v_2), the first bit of the string is set to 1, whereas it is set to 0 if a down-move (to v_3) is made. The next move, to the centre node v_4, is deterministic, and merely serves as a preparation for the next bit-generating step, which is carried out in the same way as the first move. Thus, upon reaching the end of the chain, the artificial ant will have generated a

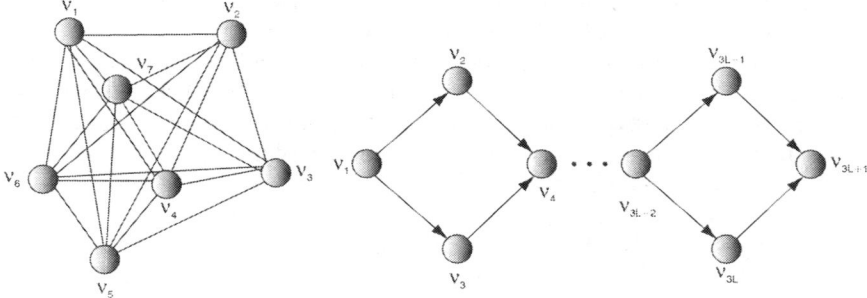

Figure 4.5: *Construction graphs for ACO. Left panel: a typical construction graph for the TSP. Right panel: a chain construction graph.*

string consisting of L bits. Chain construction graphs will be considered further in connection with our discussion of some of the mathematical properties of ACO algorithms in Appendix B. Let us now return to the TSP, and introduce ACO in detail. Since the introduction of ACO by Dorigo [17, 18], several different versions have been developed. Here, we will consider two versions, namely Ant system (AS) and Max–min ant system (MMAS), using the special case of the TSP in the description of the algorithms.

4.2.1 Ant system

AS was the first ACO algorithm introduced [18]. In order to solve TSP over n nodes, a set of N artificial ants are generated and released onto the construction graph $\mathcal{G}$. The formation of a candidate solution to the problem involves traversing a path that will take an ant to each city (node) once, and then return to the origin. Thus, each ant starts with an empty candidate solution $S = \emptyset$ and, for each move, an element representing the index of the current node is added to S. In order to ascertain that each node is visited only once, each ant maintains a memory in the form of a **tabu list** $L_T(S)$, containing the (indices of the) nodes already visited. Thus, initially, the node of origin is added to L_T. As mentioned above, each edge e_{ij} of the graph is associated with a pheromone level τ_{ij}. In any given step, an ant chooses its move probabilistically, based on a factor depending on the pheromone level as well as the distances between the current node v_j and the potential target nodes. More specifically, the probability of the ant taking a step from node j to node i is given by

$$p(e_{ij}|S) = \frac{\tau_{ij}^{\alpha} \eta_{ij}^{\beta}}{\sum_{v_l \notin L_T(S)} \tau_{lj}^{\alpha} \eta_{lj}^{\beta}}, \tag{4.3}$$

where, in the TSP, $\eta_{ij} = 1/d_{ij}$ and d_{ij} is the (Euclidean) distance between node j and node i. α and β are constants that determine the relative importance (from the point of view of the artificial ants) of the pheromone trails and the problem-specific information η_{ij}, referred to as the **visibility** in the case of TSP.

In each step of the movement, the current node is added to the tabu list L_T. When all nodes but one have been visited, only the as yet unvisited node is absent from the tabu list, and the move to that node therefore occurs with probability 1, as obtained from eqn (4.3). Next, the tour is completed by a return to the city of origin (which is obtained as the first element of L_T).

eqn (4.3) specifies, albeit probabilistically, the movement of the ants, given the pheromone levels τ_{ij}. However, the equation says nothing about the variation of pheromone levels. In order to complete the description of the algorithm, this crucial element must also be introduced. In AS, an **iteration** consists of the evaluation of all N ants. At the end of each iteration, all ants contribute to the pheromone levels on the edges of the construction graph. Let D_k denote the length of the tour generated by ant k. The pheromone level on edge e_{ij} is then increased by ant k according to

$$\Delta \tau_{ij}^{[k]} = \begin{cases} \frac{1}{D_k} & \text{if ant } k \text{ traversed the edge } e_{ij}, \\ 0 & \text{otherwise.} \end{cases} \qquad (4.4)$$

The total increase in the pheromone level on edge e_{ij} is thus given by

$$\Delta \tau_{ij} = \sum_{k=1}^{N} \Delta \tau_{ij}^{[k]}. \qquad (4.5)$$

Obviously, with eqn (4.5), pheromone levels will increase indefinitely. However, taking a cue from the properties of real pheromone trails, AS introduces pheromone evaporation, and the complete update rule thus takes the form

$$\tau_{ij} \leftarrow (1-\rho)\tau_{ij} + \Delta \tau_{ij}, \qquad (4.6)$$

where $\rho \in \,]0,1]$ is the **evaporation rate**. Note that eqn (4.6) is applied to *all* edges in the construction graph. Hence, for edges e_{ij} that are not traversed by any ant, the pheromone update consists only of evaporation. AS is summarized in Algorithm 4.1.

The free parameters in AS are the constants α, β and ρ, as well as N (the number of ants). The optimal values of these parameters will, of course, vary between problems. However, experimental studies have shown that, over a large range of problems, the values $\alpha = 1$, $\beta \in [2,5]$, $\rho = 0.5$ and $N = n$ (i.e. the number of nodes in the construction graph) yield good performance. A remaining issue is the initialization of the pheromone trails. In AS, it is common to initialize the pheromone levels as $\tau_{ij} = \tau_0 = N/D^{\text{nn}}$, where D^{nn} is the **nearest-neighbour tour** obtained by starting at a random node and, at each step, moving to the nearest unvisited node.

1. Initialize pheromone levels:

   $\tau_{ij} = \tau_0, \quad \forall i, j \in [1, n].$

2. For each ant k, select a random starting node, and add it to the (initially empty) tabu list L_T. Next, build the tour S. In each step of the tour, select the move from node j to node i with probability $p(e_{ij}|S)$, given by:

   $$p(e_{ij}|S) = \frac{\tau_{ij}^\alpha \eta_{ij}^\beta}{\sum_{v_l \notin L_T(S)} \tau_{lj}^\alpha \eta_{lj}^\beta}.$$

   In the final step, return to the node of origin, i.e. the first element in L_T. Finally, compute and store the length D_k of the tour.

3. Update the pheromone levels:
   3.1. For each ant k, determine $\Delta \tau_{ij}^{[k]}$ as:

   $$\Delta \tau_{ij}^{[k]} = \begin{cases} \frac{1}{D_k} & \text{if ant } k \text{ traversed the edge } e_{ij}, \\ 0 & \text{otherwise.} \end{cases}$$

   3.2. Sum the $\Delta \tau_{ij}^{[k]}$ to generate $\Delta \tau_{ij}$:

   $$\Delta \tau_{ij} = \sum_{k=1}^{N} \Delta \tau_{ij}^{[k]}.$$

   3.3. Modify τ_{ij}:

   $$\tau_{ij} \leftarrow (1 - \rho)\tau_{ij} + \Delta \tau_{ij}.$$

4. Repeat steps 2 and 3 until a satisfactory solution has been found.

Algorithm 4.1: *Ant system (AS), applied to the case of TSP. See the main text for a complete description of the algorithm.*

Example 4.2
Algorithm 4.1 was implemented in a computer program, and was then applied to the TSP problem illustrated in Fig. 4.6. In this instance of TSP, there were 50 cities to visit. Several parameter settings were tried. For each parameter setting, 100 runs were carried out. The number of ants was equal to 50, and each run lasted 200 iterations, so that the number of evaluated paths was equal to 10,000 for each run. The results are summarized in Table 4.1. The distribution of results for any given parameter setting could be approximated rather well

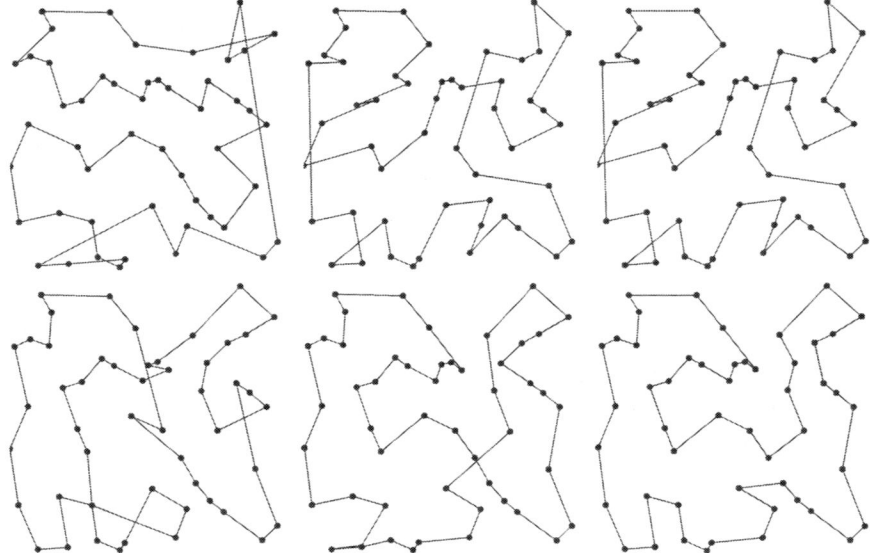

Figure 4.6: *An instance of the TSP, considered in Examples 4.2 and 4.3. The panels show the best paths found in iteration 1 (upper left), 2 (upper middle), 3 (upper right), 5 (lower left), 10 (lower middle) and 100 (lower right) of a typical ACO run.*

Table 4.1: *Results obtained by applying AS to the instance of TSP shown in Fig. 4.6.*

α	β	ρ	τ_0	$\overline{D}$	s
1	2	0.5	0.3	132.94	1.12
1	2	0.5	0.1	132.95	1.11
1	5	0.5	0.1	132.50	1.03
1	2	0.8	0.1	133.31	1.70
1	2	0.2	0.1	133.14	1.03
1	2	0.5	0.01	132.90	1.16

The average path length $\overline{D}$ and estimated standard deviation s were based on 100 runs.

by a normal distribution. The table shows both the averages over the 100 runs, as well as the estimated standard deviation, obtained using eqn (C9). As is evident from the table, the algorithm is rather insensitive to the choice of parameters, in all cases achieving a result fairly close to the optimal one (with length $D^* = 127.28$). ■

4.2.2 Max–min ant system

MMAS, introduced in Ref. [70], is a version of AS that attempts to exploit good candidate solutions more strongly than AS. While the MMAS algorithm shares many features with AS, there are also important differences between the two algorithms. First of all, in MMAS, only the ant generating the best solution is allowed to deposit pheromone. The *best solution* can either be defined as the *best in the current iteration* or the *best so far*. In the latter case, the *best so far* solution must, of course, be stored and updated whenever a new, better solution is found. Thus, continuing with the TSP as an example, letting D_b denote the length of the current best tour, in MMAS the pheromone levels change according to

$$\Delta \tau_{ij} = \Delta \tau_{ij}^{[b]}, \qquad (4.7)$$

where $\tau_{ij}^{[b]} = 1/D_b$ if the best ant traversed the edge e_{ij} in its tour, and 0 otherwise. Note that, as in AS, pheromone levels are updated for *all* edges, according to eqn (4.6), even though only n edges receive a positive contribution according to eqn (4.7). Thus, the edges that are not in the tour of the best ant are subject only to pheromone evaporation.

Second, since the strong exploitation of good solutions may lead to stagnation, where all ants essentially follow the same trail, in MMAS limits are introduced on the pheromone levels. Thus, after the pheromone update according to eqns (4.6) and (4.7), the levels are modified such that if $\tau_{ij} > \tau_{max}$ then $\tau_{ij} \leftarrow \tau_{max}$ and, likewise, if $\tau_{ij} < \tau_{min}$, then $\tau_{ij} \leftarrow \tau_{min}$. The introduction of the lower limit τ_{min} implies a lower bound on the probability of traversing any edge e_{ij}. This lower bound makes it possible to prove a convergence theorem for MMAS, see Appendix B, Section B.3.2. Furthermore, in the simple case of the Onemax problem (see Chapter 3), one can also derive an upper bound for the estimated runtime of the algorithm, albeit with some additional simplifications. Such an estimate can be found in Appendix B, Section B.3.3. A third difference between MMAS and AS concerns the pheromone initialization. In MMAS, pheromone levels are initialized to the maximum value τ_{max} so that, initially,

$$\tau_{ij} = \tau_{max} \quad \forall\, i, j \in [1, n]. \qquad (4.8)$$

With this initial setting, there will be a greater degree of exploration in the early stages of the algorithm, before pheromone levels start dropping on edges that define unfavourable (i.e. long) tours.

The choice of τ_{max} is based on estimates of the theoretical upper limit of the pheromone levels under the update rule given in eqns (4.6) and (4.7). In fact, it can easily be shown (see Appendix B, Section B.3.1) that an upper bound of the pheromone deposited on any edge e_{ij} is given by $1/(\rho D^*)$, where D^* is the length of the optimal tour.[3] Obviously, during the optimization procedure, D^* is not yet

---

[3] More generally, the upper bound can be written f^*/ρ, where f^* is the value of the objective function for the optimal solution, which equals $1/D^*$ in the case of the TSP.

known. Therefore, $\tau_{\max}$ is set to $1/(\rho D_b)$, where D_b again denotes the length of the current best tour. Initially one may, for example, take $\tau_{ij} = 1/(\rho D^{nn})$, where D^{nn} is the nearest-neighbour tour described in connection with AS above. Whenever a new best tour is found, $\tau_{\max}$ is updated using the new value of D_b. Thus, in MMAS, the value of $\tau_{\max}$ changes dynamically when the algorithm is running. $\tau_{\min}$ is often chosen through experimentation. Using a combination of analytical and experimental arguments, a suitable value for $\tau_{\min}$ can be obtained (for TSP, with $n \gg 1$) as [70]

$$\tau_{\min} = \frac{\tau_{\max}(1 - \sqrt[n]{0.05})}{(n/2 - 1)\sqrt[n]{0.05}}. \tag{4.9}$$

---

1. Initialize pheromone levels:

    $\tau_{ij} = \tau_{\max}, \quad \forall\, i,j \in [1,n]$.

2. For each ant k, select a random starting node, and add it to the (initially empty) tabu list L_T. Next, build the tour S. In each step of the tour, select the move from node j to node i with probability $p(e_{ij}|S)$, given by:

    $$p(e_{ij}|S) = \frac{\tau_{ij}^{\alpha} \eta_{ij}^{\beta}}{\sum_{v_l \notin L_T(S)} \tau_{lj}^{\alpha} \eta_{lj}^{\beta}}.$$

    In the final step, return to the node of origin, i.e. the first element in L_T. Finally, compute and store the length D_k of the tour.

3. Update the pheromone levels:

    3.1. For the best ant (either the *best in the current iteration* or the *best so far*) determine $\Delta \tau_{ij}^{[b]}$ as:

    $$\Delta \tau_{ij}^{[b]} = \begin{cases} \frac{1}{D_b} & \text{if the best ant traversed the edge } e_{ij}, \\ 0 & \text{otherwise.} \end{cases}$$

    3.2. Modify τ_{ij}:

    $$\tau_{ij} \leftarrow (1 - \rho)\tau_{ij} + \Delta \tau_{ij}^{[b]}.$$

    3.3. Impose pheromone limits, such that $\tau_{\min} \leq \tau_{ij} \leq \tau_{\max}$, $\forall\, i,j \in [1,n]$.
    3.4. If $1/(\rho D_b) > \tau_{\max}$ then update $\tau_{\max}$ as $\tau_{\max} \leftarrow 1/(\rho D_b)$.

4. Repeat steps 2 and 3 until a satisfactory solution has been found.

Algorithm 4.2: *Max–min ant system (MMAS), applied to the case of the TSP. See the main text for a complete description of the algorithm.*

Table 4.2: *Results obtained applying MMAS to the instance of the TSP shown in Fig. 4.6.*

α	β	ρ	τ_{max}	τ_{min}	$\overline{D}$	s
1	2	0.5	0.0133	3.43×10^{-5}	136.11	3.38
1	5	0.5	0.0133	3.43×10^{-5}	128.87	1.69
1	2	0.8	0.0083	2.14×10^{-5}	129.19	1.82
1	2	0.2	0.0333	8.57×10^{-5}	128.66	1.57

The average $\overline{D}$ and estimated standard deviation s were based on 100 runs.

Despite the pheromone limits, it is not uncommon that the MMAS algorithm gets stuck on a suboptimal solution. Thus, a useful approach is to restart the algorithm, using the most recent available estimate of D^* (i.e. the current D_b) to determine τ_{max}, whenever no improvements are detected over a number of iterations.

It should also be noted that, at least for TSP with large (>200) values of n, better results are obtained by alternating the definition of the best solution. Thus, the *best in the current iteration* is used when updating the pheromones for a number of iterations, after which a switch is made to the *best so far* etc. The MMAS algorithm is summarized in Algorithm 4.2.

Example 4.3
Algorithm 4.2 was implemented in a computer program. The pheromone updates were based on the *best so far* solution, found in the current run. The upper pheromone limit τ_{max} was initially set as $1/(\rho D^{nn})$, with $D^{nn} \approx 150$, and the lower pheromone limit τ_{min} was set according to eqn (4.9). τ_{max} was then gradually adjusted, as described above.

The program was applied to the same instance of the TSP considered in Example 4.2, using the same parameter settings, except for the initialization of the pheromone levels. Again, 100 runs were carried out for each parameter setting. The results are summarized in Table 4.2. As can be seen, the distribution of results is generally wider than for AS, but the average results obtained with MMAS are, in several cases, superior to those obtained using AS. ■

4.3 Applications

In view of their biological background, it is not surprising that ACO algorithms have been applied to a variety of problems involving routing. The TSP, described above, is perhaps the foremost example, but ACO algorithms have also been used for solving other routing problems involving, for example, telecommunication networks [68]. Here, however, we shall consider another application, namely machine scheduling.

4.3.1 Single-machine scheduling

In the general problem of **job shop scheduling**, n jobs, consisting of a finite sequence of operations, are to be processed on m machines, in the shortest possible time, while fulfilling constraints regarding the order of precedence between the different operations in the jobs. This is an example of a **scheduling problem**, for which ACO algorithms have been developed [3, 14]. Here we shall consider another scheduling problem, namely the **single-machine total weighted tardiness problem** (SMTWTP). In this problem, the aim is to schedule the order of execution of n jobs, assigned to a single machine. Let t_j denote the processing time of job j, and d_j its due date (i.e. the deadline). The tardiness T_j of a job j is defined as

$$T_j = \max(0, c_j - d_j), \qquad (4.10)$$

where c_j is the actual completion time of the job. Thus, jobs that are completed on time receive a tardiness of 0 whereas, for jobs that are not completed on time, the tardiness equals the delay. The aim of the single-machine total tardiness problem (SMTTP) is to minimize the total tardiness T_{tot}, given by

$$T_{\text{tot}} = \sum_{j=1}^{n} T_j, \qquad (4.11)$$

whereas, in the weighted problem (SMTWTP), the goal is to minimize the weighted total tardiness, defined as

$$T_{\text{tot,w}} = \sum_{j=1}^{n} w_j T_j, \qquad (4.12)$$

for some pre-specified values of the weights $w_j \geq 0, j = 1, \ldots, n$. In this problem, any permutation of the n jobs is a feasible solution. Thus, in order to generate a solution using an ACO algorithm, the ants must build a sequence of jobs, such that each job is executed exactly once. As in TSP, the ants base their choice on a combination of two factors: the pheromone level τ_{ij} involved in placing job i immediately after job j, and a problem-specific factor η_{ij}, see eqn (4.3). In the TSP, the latter factor involved the inverse of the distance between the nodes j and i. In SMT(W)TP, by contrast, the problem-specific factor involves a measure of time, instead of distance. Several different measures have been used in the literature, e.g.

$$\eta_{ij} = \frac{1}{t_j}, \qquad (4.13)$$

a measure referred to as the **shortest processing time** (SPT), which is efficient in cases where most jobs are late. Other measures involve the due date; the simplest such measure

$$\eta_{ij} = \frac{1}{d_j}, \qquad (4.14)$$

is referred to as **earliest due date** (EDD). The two measures above are obviously simple to obtain, but may not be the most efficient, since they do not take into account the growing sequence of scheduled jobs. An alternative is to use the **modified due date** (MDD), defined as

$$\eta_{ij} = \frac{1}{\max(T^{[s]} + t_j, d_j)}, \tag{4.15}$$

where $T^{[s]}$ denotes the total processing time of the jobs already scheduled. In order to use this measure, the factors η_{ij} must, of course, be updated after each addition to the job sequence.

Using any of the measures η_{ij} above, an ACO algorithm (e.g. AS, as described above) can easily be set up for solving the SMT(W)TP. However, it should be noted that, for this particular problem, it is common to use another version of ACO, known as **Ant-colony system** (ACS) which, however, is beyond the scope of this text. In addition, the candidate solutions to the SMT(W)TP obtained by the artificial ants are usually enhanced by **local search**. This procedure, which can be applied either to all candidate solutions in the current iteration or (to reduce the amount of computation) only to the best candidate solution of the current iteration, involves searching for neighbouring candidate solutions, using e.g. **interchange**, in which all $n(n-1)/2$ interchanges of pairs of jobs are considered. Thus, for example, if the job sequence is $(1, 6, 4, 7, 3, 5, 2)$ (in a simple case involving only seven jobs), an interchange of the second and fifth jobs in the sequence results in $(1, 3, 4, 7, 6, 5, 2)$. An alternative procedure is **insertion**, in which the job at position i_1 in the sequence is extracted and reinserted in position i_2. Thus, again considering the sequence $(1, 6, 4, 7, 3, 5, 2)$, extraction of the second job, followed by reinsertion in the fifth position, results in the sequence $(1, 4, 7, 3, 6, 5, 2)$. Obviously, the two processes of interchange and insertion can also be combined.

In cases with small n (up to around 50), classical integer programming techniques, such as the **branch-and-bound** method can be counted on to solve the SMT(W)TP. However, for larger n, ACO-based methods become useful. It should be noted that the level of difficulty in solving randomly generated instances of SMT(W)TP is strongly variable: for some instances of the problem the best solution can be found quite easily whereas others can be very challenging. Thus, in order to investigate performance, a given method must be tested against multiple instances of the problem. A procedure for generating instances of SMTTP is given in Ref. [3]. Here, the processing times t_j for the n jobs are integers chosen from the uniform distribution in the range $[1, n]$, and the due dates d_j are also integers taken from the uniform distribution in the range D given by

$$D = \left[\sum_{j=1}^{n} t_j \left(1 - c_1 - \frac{c_2}{2}\right), \sum_{j=1}^{n} t_j \left(1 - c_1 + \frac{c_2}{2}\right) \right], \tag{4.16}$$

where c_1 is the **tardiness factor** and c_2 is the **relative range of due dates**. The complexity of the problem varies with c_1 and c_2. In Ref. [3], these parameters were both taken from the set $\{0.2, 0.4, 0.6, 0.8, 1.0\}$. For each of the $5 \times 5 = 25$ possible combinations, five instances of the problem were generated for $n = 50$ and $n = 100$. Using ACO the optimal (or, rather, best known) solution was found in 124 out of 125 cases for $n = 50$, and in all 125 cases for $n = 100$. In Ref. [14], similar results were reported for the SMTWTP. These two studies show that, for the SMT(W)TP, ACO algorithms generally outperform other stochastic methods.

4.3.2 Co-operative transport using autonomous robots

As mentioned in the introduction to this chapter, many species of ants are capable of co-operative transport of objects, a process that has inspired robotics researchers to investigate co-operative behaviours in autonomous robots. Ants participating in co-operative transport appear to co-ordinate their motion based on stigmergy, i.e. indirect communication, something that also makes the procedure suitable for robotic applications by avoiding the problem of direct communication between the participating robots. In fact, however, little is known about the detailed processes involved in ant movement coordination, and therefore cooperative transport may turn out to be an instance where the connection between biology and robotics goes both ways, i.e. where robotics may also inspire further biological research. Thus, even though the following example is not, strictly speaking an *application* of ACO, it is strongly related to such algorithms.

Co-operative transport using autonomous robots has been studied by, among others, Kube and Zhang [40], using both simulations and actual robots. In their experiments, a group of identical autonomous (wheeled) robots were given the task of locating a box near the centre of an arena and then co-operatively push the box to the edge of the arena. Of course, the box was too heavy to be pushed by a single robot: in order for the box to move, a minimum of two robots, pushing in roughly the same direction, was required. The problem is illustrated schematically in the left panel of Fig. 4.7.

The problem was solved in the framework of **behaviour-based robotics**, in which the control system of a robot is built in a bottom-up fashion from a set of elementary behaviours. Starting with simulation experiments, Kube and Zhang developed a robotic control system consisting of five elementary behaviours, namely *Find* (for locating the box), *Slow* (for slowing down, so as to avoid rear-end collisions with another robot), *Follow* (for following another robot), *Avoid* (for avoiding collisions) and *Goal* (for directing a robot towards the box). In addition, a structure for **behaviour selection**, i.e. for timely activation of the various behaviours, was generated. A full description of this mechanism will not be given here. It suffices to say that the experiments were successful, resulting in reliable co-operative transport of the box (in a more or less random direction, depending on the exact distribution of robots around the box).

Building upon the lessons learned in the simulation experiment, Kube and Zhang were also able to transfer their results successfully to real robots, as illustrated in

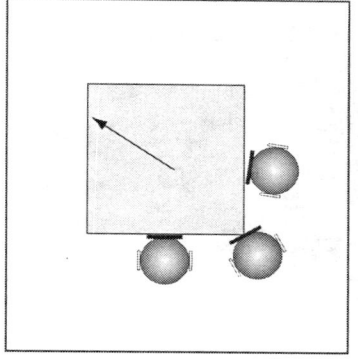 

Figure 4.7: *Left panel: schematic illustration of a box-pushing task, in which three wheeled robots are attempting to move a large, rectangular box towards the edge of the arena. The arrow indicates the approximate instantaneous direction of motion. Right panel: five robots co-operating on a box-pushing task. Copyright Ronald Kube. Reproduced with permission.*

the right panel of Fig. 4.7. The robots were equipped with sensors capable of detecting the goal, which was illuminated by a light bulb, as well as obstacles in the environment. Furthermore, the problem of deadlocks was considered. Each robot was equipped with a counter, measuring the time t elapsed since the latest move. In case the box got stuck, such that the robots kept pushing without generating any movement, t would, of course, eventually exceed a threshold T, at which point the robot in question would change the direction of the applied force. If this also failed, repositioning was tried, i.e. the robot would move to a different location around the perimeter of the box, and continue pushing.

Even though this simple box-pushing task may not appear to be very relevant at first glance, the mere fact that successful co-operative transport could be generated is important, as it paves the way for more complex, real-world applications, for example in rescue missions. In such cases, a swarm of robots would be released in a disaster area (resulting, for example, from an earthquake or a fire) to search efficiently for disaster victims. A robot discovering an injured person would then recruit other robots to the scene, and the group of robots would co-operatively move the person to safety.

In addition to the experiments described by Kube and Zhang, other studies of ant-inspired co-operative robot behaviours have been carried out within the framework of the **Swarm-bot project** [51]. This project involved the development of a fairly simple, cost-effective robot (the **s-Bot**), that could be mass-produced for use in multi-robot applications. Figure 4.8 shows some s-Bots in action. Several tasks were considered within the framework of the Swarm-bot project. One such task, namely dynamic bridge-building in order to pass a large obstacle, is illustrated in the right panel of Fig. 4.8.

Figure 4.8: *Left panel: two s-Bots. The s-Bot on the right is using its gripper to grasp the s-Bot on the left. Right panel: a swarm-bot made of five connected s-Bots passing a step. Copyright Marco Dorigo. Reprinted with permission.*

Exercises

4.1 Implement the ant navigation model described in Section 4.1 and repeat the experiments described in Example 4.1 for different values of m, C and r.

4.2 Assuming that the results in Table 4.1 follow normal distributions, can the null hypothesis (equal performance of all parameter sets) be rejected for any pair of parameter sets? (See Appendix C).

4.3 Write a computer program implementing AS (Algorithm 4.1). Next, apply the program to an instance of TSP where the coordinates $(x_{1i}, x_{2i})^T$ of node v_i, $i = 1, \ldots, n$ are given by

$$x_{1i} = 0.1((9 + 11i^2) \bmod 200), \tag{4.17}$$

$$x_{2i} = 0.1((7 + 1723i) \bmod 200). \tag{4.18}$$

Solve the problem for $n = 30, 50$ and 100. What is the length of the shortest path found for each value of n?

4.4 Repeat Exercise 4.3, using MMAS. Unlike Example 4.3, let the definition of the best solution vary between *best in current iteration* and *best so far* every k^{th} iteration. How do the results of the MMAS program vary with k?

4.5 Implement the simplified ACO described in connection with the runtime estimate in Appendix B, Section B.3.3, and apply it to the Onemax problem. Compare the results found with those obtained using the estimate given in eqn (B75).

Chapter 5

Particle swarm optimization

5.1 Biological background

The tendency to form swarms[1] appears in many different organisms, for instance, (some species of) birds and fish. Swarming behaviour offers several advantages, for example protection from predators: an animal near the centre of a swarm is unlikely to be captured by a predator. Furthermore, the members of a swarm may confuse predators through coordinated movements, such as rapid division into subgroups. Also, because of the large concentration of individuals in a swarm, the risk of injury to a predator, should it attack, can be much larger than when it is attacking a single animal. On the other hand, it is not entirely evident that swarming is always beneficial: once a predator has discovered a swarm, at least it knows where the prey is located. However, by aggregating, the prey effectively presents the predator with a needle-in-haystack problem and the benefits of doing so are apparently significant, since swarming behaviour is so prevalent in nature. Swarms are also formed for other reasons than protection from predators. For example, swarming plays a role in efficient reproduction: it is easier to find a mate in a large group. However, also in this case, there are both benefits and drawbacks, since competition obviously increases as well. Swarming is also a prerequisite for cooperation that, in turn, plays a crucial role in, for example, the foraging (food gathering) of several species, particularly ants, bees and termites, but also in other organisms such as birds. Here, the principle is that many eyes are more likely to find food than a single pair of eyes.

---

[1] Even though, normally, several different terms are used to describe different kinds of swarms, e.g. schools of fish, flocks of birds, herds of buffalo, etc., in this chapter the generic term *swarm* will be used throughout.

The search efficiency provided by swarming is what underlies **particle swarm optimization** (PSO) algorithms. Before introducing PSO in detail, however, we shall briefly consider a model for swarming in biological organisms.

5.1.1 A model of swarming

In many instances of swarming in animals, there is no apparent leader that the other members of the swarm follow. Thus, the swarm must be a result of *local* interactions only, and an interesting question follows: how can a coherent swarm result from such interactions? This issue has been addressed by Reynolds [62], who introduced a model for numerical simulation of the swarming of bird-like objects (or **boids** as they were called), which we will consider next.

In the description of PSO below, the ith member of a swarm, referred to as a *particle*, will be denoted p_i. Thus, in order to keep a unified notation throughout the chapter, we shall use the symbol p_i also for the boids considered here. Let **S**, defined as,

$$S = \{p_i, \quad i = 1, \ldots, N\}. \tag{5.1}$$

denote a swarm of N boids. In this model, the boids have limited visual range and can therefore only perceive nearby swarm mates. Thus, for each boid i, a **visibility sphere** $\mathbf{V}_i$ is introduced, defined as

$$\mathbf{V}_i = \{p_j : \|\mathbf{x}_j - \mathbf{x}_i\| < r, \quad j \neq i\}, \tag{5.2}$$

where r is a global constant, i.e. a property of the swarm. Note that vision is isotropic in this basic model. Of course, the model can easily be modified to include directionality of vision. In each time step of the simulation, the positions $\mathbf{x}_i$ and velocities $\mathbf{v}_i$ of the boids are updated using standard Euler integration, i.e.

$$\mathbf{v}_i \leftarrow \mathbf{v}_i + \mathbf{a}_i \Delta t, \quad i = 1, \ldots, N, \tag{5.3}$$

$$\mathbf{x}_i \leftarrow \mathbf{x}_i + \mathbf{v}_i \Delta t, \quad i = 1, \ldots, N, \tag{5.4}$$

where $\mathbf{v}_i$ denotes the velocity of boid i, $\mathbf{a}_i$ its acceleration, and Δt is the time step. Each boid is influenced by three different movement tendencies (or steers) that together determine its acceleration, namely **cohesion, alignment** and **separation**. Cohesion represents the tendency of any given boid to stay near the centre of the swarm. Let ρ_i denote the centre of density of the boids within the visibility sphere of boid i, i.e.

$$\rho_i = \frac{1}{k_i} \sum_{p_j \in V_i} \mathbf{x}_j, \tag{5.5}$$

where k_i is the number of boids in $\mathbf{V}_i$. The steering vector representing cohesion is defined as

$$\mathbf{c}_i = \frac{1}{T^2}(\rho_i - \mathbf{x}_i), \tag{5.6}$$

where T is a time constant, introduced in order to give $\mathbf{c}_i$ the correct dimension (acceleration). If $k_i = 0$, i.e. if no boids are within the visibility sphere of boid i, then $\mathbf{c}_i$ is set to $\mathbf{0}$. Alignment, by contrast, is the tendency of boids to align their velocities with those of their nearby swarm mates. Thus, the alignment steering vector is defined as

$$\mathbf{l}_i = \frac{1}{Tk_i} \sum_{p_j \in V_i} \mathbf{v}_j. \tag{5.7}$$

As in the case of cohesion, if $k_i = 0$, then $\mathbf{l}_i$ is set to $\mathbf{0}$. Separation, finally, is needed in order to avoid collision with nearby boids, and the corresponding steering vector is obtained as

$$\mathbf{s}_i = \frac{1}{T^2} \sum_{p_j \in V_i} (\mathbf{x}_i - \mathbf{x}_j). \tag{5.8}$$

If $k_i = 0$, then $\mathbf{s}_i = \mathbf{0}$. Finally, combining the steering vectors, the acceleration of boid i is obtained as

$$\mathbf{a}_i = C_c \mathbf{c}_i + C_l \mathbf{l}_i + C_s \mathbf{s}_i, \tag{5.9}$$

where the constants C_c, C_l and C_s ($\in [0, 1]$) measure the relative impact of the three steering vectors. An example of steering vectors in a boids simulation is shown in Fig. 5.1.

A crucial factor in simulations based on this model is the initialization; if the initial speed of the boids is too large, the swarm will break apart. Thus, commonly, swarms are instead initialized with $\mathbf{v}_i \approx \mathbf{0}$, $i = 1, \ldots, N$. Furthermore, each boid should (at initialization) be within the visibility sphere of at least one other boid.

The simple model presented above leads, in fact, to very realistic swarm behaviour, and the algorithm has (with some modifications) been used both in computer games and in movies (e.g. Jurassic Park). An illustration of the boids model is given in Example 5.1.

Example 5.1
The boids model was implemented in a computer program, with three-dimensional visualization of the swarm. Even though the equations described above do generate swarm behaviour in principle, the equations must, in practice, be supplemented with limits on accelerations in order for the swarm to remain coherent. Here, the magnitude of each steer was limited to $a_{\max}$. Furthermore, due to random fluctuations, a swarm is likely, over time, to drift away indefinitely. Thus, in order to keep the swarm in roughly the same area, an additional acceleration component defined as

$$\mathbf{e}_i = \begin{cases} -\frac{1}{T^2} \mathbf{x}_i & \text{if } \|\mathbf{x}_i\| > R_{\max}, \\ 0 & \text{otherwise}, \end{cases} \tag{5.10}$$

was added to $\mathbf{a}_i$. $R_{\max}$ is a constant determining (roughly) the maximum allowed distance, from the origin, of any boid. Finally, the overall velocities of the boids were limited as $\|\mathbf{v}_i\| < v_{\max}$, $i = 1, \ldots, N$.

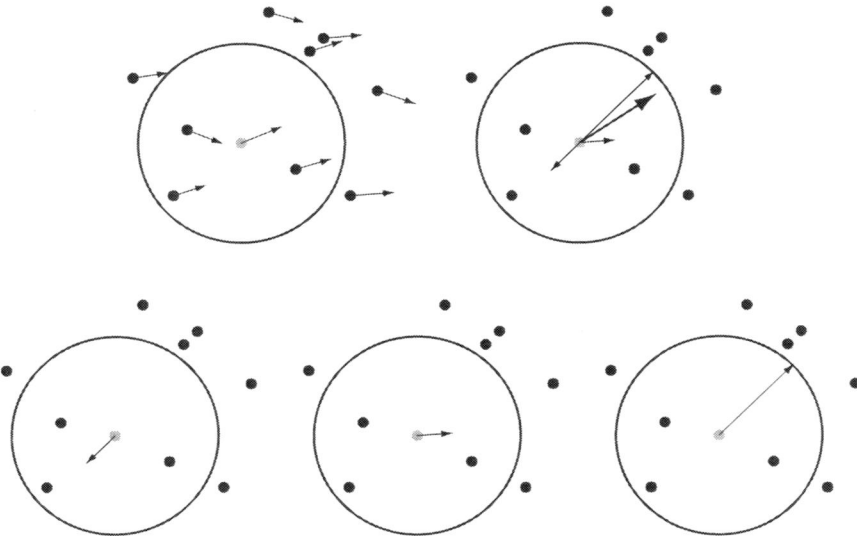

Figure 5.1: *A two-dimensional example of steering vectors: The upper left panel shows the positions and velocities of a swarm of boids, and the upper right panel shows the resulting acceleration vector (thick arrow) for the boid in the centre, as well as its components* **c**, **l** *and* **s**. *The lower panel shows, from left to right, the individual components* **c**, **l** *and* **s**, *respectively. In this example, the adjustable parameters were set to* $T = C_c = C_l = C_s = 1$. *The circles indicate the size of the visibility sphere.*

Several runs were made, with different parameter values. In one of the best runs, depicted in Fig. 5.2, the number of boids was set to 50. At initialization, the boids were randomly distributed in a sphere of radius 2.5, and given random velocities such that $\|\mathbf{v}_i\| < 0.03$, $i = 1, \ldots, N$. The radius of the visibility sphere (r) was set to 2.0, and the time constant T was set to 5.0. The constants C_c, C_l and C_s were set to 1.0, 0.1 and 0.75, respectively. Finally $a_{\max}$, $v_{\max}$ and $R_{\max}$ were set to 0.03, 0.15 and 6.0, respectively. ∎

5.2 Algorithm

Particle swarm optimization (PSO) [36] is based on the properties of swarms. As is the case with ACO, PSO algorithms (of which there are several versions, as we shall see) attempt to capture those aspects of swarming that are important in optimization, namely, the search efficiency attributable to a swarm. Essentially, in PSO, each particle[2] is associated both with a position and a velocity in the search

---

[2] As mentioned earlier in the chapter, in PSO, the candidate solutions, corresponding to individuals in EAs, are referred to as *particles*.

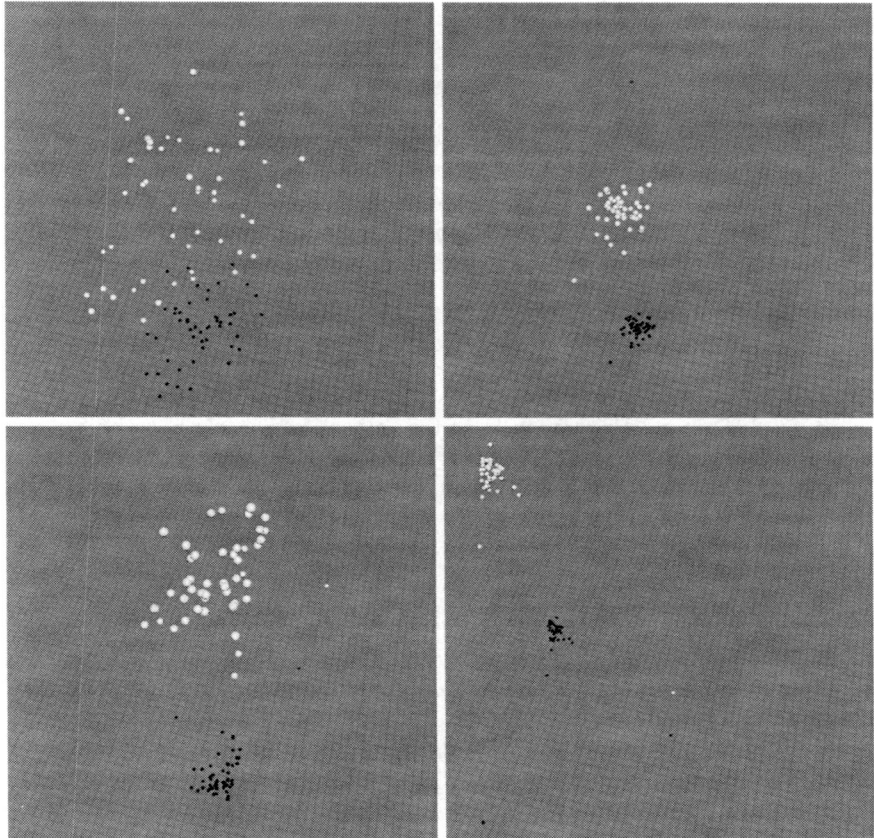

Figure 5.2: *Four snapshots from the boids simulation described in Example 5.1. The upper left panel shows the initial distribution of the boids, and the upper right panel shows the results of the early contraction that occurred since the magnitudes of the initial velocities were quite small. The lower panels depict the swarm at two later instances, showing that it remains coherent, with the exception of a few outliers.*

space, as well as a method for determining the changes in velocity depending on the performance of the particle itself and that of other particles. Thus, a clear difference compared, for example, with EAs is the introduction of a velocity in the search space. A basic PSO algorithm is described in Algorithm 5.1. The first step is initialization of the position $\mathbf{x}_i$ and the velocity $\mathbf{v}_i$ of each particle $p_i, i = 1, \ldots, N$. The appropriate number of particles will vary from problem to problem, but is typically smaller than the number of individuals used in EAs. Common values of N are 20–40. Positions are normally initialized randomly, using uniform sampling in a given range $[x_{\min}, x_{\max}]$, i.e.

$$x_{ij} = x_{\min} + r(x_{\max} - x_{\min}), \quad i = 1, \ldots, N, \ j = 1, \ldots, n \tag{5.11}$$

where x_{ij} denotes the jth component of the position of particle p_i and r is a uniform random number in the range [0, 1]. N denotes the size of the swarm, corresponding to the population size in an EA, and n, as usual, is the dimensionality of the problem (the number of variables). Velocities are also initialized randomly, according to

$$v_{ij} = \frac{\alpha}{\Delta t}\left(-\frac{x_{\max} - x_{\min}}{2} + r(x_{\max} - x_{\min})\right), \quad i = 1, \ldots, N, \ j = 1, \ldots, n \tag{5.12}$$

where v_{ij} denotes the jth component of the velocity of particle p_i and r again is a random number in the range [0, 1]. α is a constant in the range [0, 1] (often set to 1), and Δt is the time step length which, for simplicity, commonly is set to 1. In the common special case where $x_{\min} = -x_{\max}$ eqn (5.12) is reduced to

$$v_{ij} = \alpha \frac{x_{\min} + r(x_{\max} - x_{\min})}{\Delta t}, \quad i = 1, \ldots, N, \ j = 1, \ldots, n \tag{5.13}$$

Once initialization has been completed, the next step is to evaluate the performance of each particle. As in an EA, the detailed nature of the evaluation depends, of course, on the problem at hand. Also, the sign of the inequalities in Algorithm 5.1 depends on whether the goal is to maximize or minimize the value of the objective function (here, minimization has been assumed).

Next, the velocities and positions of all particles should be updated. As the aim is to reach optimal values of the objective function, the procedure for determining velocities should, of course, keep track of the performance of the particles thus far. In fact, two such measures are stored and used in PSO, namely, the best position $\mathbf{x}_i^{\text{pb}}$ so far, of particle i and the best performance $\mathbf{x}^{\text{sb}}$ so far, of *any* particle in the swarm. Thus, after the evaluation of a particle p_i, the two performance tests described in step 3 in Algorithm 5.1 are carried out. The first test is straightforward and simply consists of comparing the performance of particle p_i with its previous best performance. The second test, however, can be carried out in different ways, depending on whether the *best performance of any particle in the swarm* is taken to refer to the *current* swarm or *all* particles considered thus far, and also depending on whether the comparison includes *all* particles of the swarm or only particles in a neighbourhood (a concept that will be further discussed below) of particle p_i. In Algorithm 5.1, it is assumed that the comparison in step 3.2 involves the whole swarm, and that the best-ever position is used as the benchmark. Thus, in this case, after the first evaluation of all particles, $\mathbf{x}^{\text{sb}}$ is set to the best position thus found. $\mathbf{x}^{\text{sb}}$ is then stored, and is updated only when the condition in step 3.2 of the algorithm is fulfilled.

1. Initialize positions and velocities of the particles p_i:

   1.1 $x_{ij} = x_{\min} + r(x_{\max} - x_{\min})$, $i = 1, \ldots, N$, $j = 1, \ldots, n$
   1.2 $v_{ij} = \frac{\alpha}{\Delta t}\left(-\frac{x_{\max} - x_{\min}}{2} + r(x_{\max} - x_{\min})\right)$, $i = 1, \ldots, N$, $j = 1, \ldots, n$

2. Evaluate each particle in the swarm, i.e. compute $f(\mathbf{x}_i)$, $i = 1, \ldots, N$.
3. Update the best position of each particle, and the global best position. Thus, for all particles p_i, $i = 1, \ldots, N$:

   3.1 if $f(\mathbf{x}_i) < f(\mathbf{x}_i^{pb})$ then $\mathbf{x}_i^{pb} \leftarrow \mathbf{x}_i$.
   3.2 if $f(\mathbf{x}_i) < f(\mathbf{x}^{sb})$ then $\mathbf{x}^{sb} \leftarrow \mathbf{x}_i$.

4. Update particle velocities and positions:

   4.1 $v_{ij} \leftarrow v_{ij} + c_1 q \left(\frac{x_{ij}^{pb} - x_{ij}}{\Delta t}\right) + c_2 r \left(\frac{x_j^{sb} - x_{ij}}{\Delta t}\right)$, $i = 1, \ldots, N$, $j = 1, \ldots, n$
   4.2 Restrict velocities, such that $|v_{ij}| < v_{\max}$.
   4.3 $x_{ij} \leftarrow x_{ij} + v_{ij} \Delta t$, $i = 1, \ldots, N$, $j = 1, \ldots, n$.

5. Return to step 2, unless the termination criterion has been reached.

Algorithm 5.1: *Basic particle swarm optimization algorithm. N denotes the number of particles in the swarm, and n denotes the number of variables in the problem under study. It has been assumed that the goal is to minimize the objective function $f(\mathbf{x})$. See the main text for a complete description of the algorithm.*

Given the current values of $\mathbf{x}_i^{pb}$ and $\mathbf{x}^{sb}$, the velocity of particle p_i is then updated according to

$$v_{ij} \leftarrow v_{ij} + c_1 q \left(\frac{x_{ij}^{pb} - x_{ij}}{\Delta t}\right) + c_2 r \left(\frac{x_j^{sb} - x_{ij}}{\Delta t}\right), \quad j = 1, \ldots, n, \quad (5.14)$$

where q and r are uniform random numbers in $[0, 1]$, and c_1 and c_2 are positive constants, typically both set to 2, so that the statistical mean of the two factors $c_1 q$ and $c_2 r$ is equal to 1. The term involving c_1 is sometimes referred to as the **cognitive component** and the term involving c_2 the **social component**. The cognitive component measures the degree of self-confidence of a particle, i.e. the degree to which it trusts its own previous performance as a guide towards obtaining better results. Similarly, the social component measures a particle's trust in the ability of the other swarm members to find better candidate solutions. Once the velocities have been updated, restriction to a given range $|v_{ij}| < v_{\max}$ is carried out, a crucial step for maintaining the coherence of the swarm, i.e. to keep it from expanding indefinitely

Table 5.1: *The average and estimated standard deviations obtained from 100 samples of the distribution of the mean of the results generated using a basic PSO, applied to the two benchmark functions Ψ_2 and Ψ_4, for different values of c_1 and c_2; see Example 5.2.*

Parameters		$\Psi_1(x_1,x_2)$		$\Psi_4(x_1,x_2,x_3,x_4)$	
c_1	c_2	Avg.	S.D.	Avg.	S.D.
2	2	3.045	0.009	21.42	2.632
1	2	3.062	0.013	26.69	3.095
2	1	3.055	0.011	24.40	2.942
1	1	3.071	0.015	25.39	3.042

(see Section 5.3.3). It should be noted that the restriction of particle velocities does *not* imply that the positions will be constrained to the range $[x_{min}, x_{max}]$; it only means that particle positions will remain *bounded*. Next, the position of particle p_i is updated as

$$x_{ij} \leftarrow x_{ij} + v_{ij}\Delta t, \quad j = 1, \ldots, n. \quad (5.15)$$

This completes the first **iteration**. Steps 2, 3, and 4 of Algorithm 5.1 are then repeated until a satisfactory solution has been found. To conclude this section, we shall now consider an example of PSO.

Example 5.2
A basic PSO algorithm was implemented as described in Algorithm 5.1, and was then applied to the problem of minimizing the benchmark functions $\Psi_1(x_1,x_2)$ and $\Psi_4(x_1,x_2,x_3,x_4)$ defined in Appendix D. Several different values of the parameters c_1 and c_2 were tested. The number of particles was set to 40, and a total of 250 iterations were carried in each run, corresponding to a total of 10,000 function evaluations. In all, 3,000 runs were made for each parameter setting, and the best results were stored. The runs were divided into 100 groups of 30 runs each. For each such group the mean of the best results was computed, resulting in 100 samples of the distribution of the mean, see Appendix C, Section C.1. Table 5.1 shows the average obtained for this distribution, along with the estimated standard deviation. From the table, one can observe that the parameter choice $c_1 = c_2 = 2$ gives the best result. Furthermore, note that this simple PSO does not do very well in the case of Ψ_4. ∎

5.3 Properties of PSO

As in the case of EAs and ACO algorithms, there exist many variations on the theme provided by the basic PSO algorithm described in Algorithm 5.1, some of which will now be described.

5.3.1 Best-in-current-swarm vs. best-ever

A crucial component in PSO is the concept of the best-in-swarm performance, i.e. $\mathbf{x}^{sb}$ as introduced in Algorithm 5.1 above. The first modification we will consider concerns the scope of the comparison with respect to the iterations carried out during optimization. In Algorithm 5.1, we determined $\mathbf{x}^{sb}$ as the best-ever position $\mathbf{x}^{sb,e}$ of any particle of the swarm. An alternative approach is to consider only the best position $\mathbf{x}^{sb,c}$ in the *current* swarm, i.e. among the N particles forming the current iteration. Note that, apart from the determination of the best-in-swarm position, all other steps are identical to those of Algorithm 5.1. In terms of computer programming, the only difference between the two methods is the single line of code needed to reset the best-in-swarm in each iteration.

5.3.2 Neighbourhood topologies

In the boids model presented in the beginning of this chapter, a visibility sphere was associated with each boid, and only those boids that happened to be inside this sphere influenced the acceleration of the boid under consideration. A similar idea (albeit with an important difference, see below) has been introduced in connection with PSO, namely the concept of **neighbourhoods**.

Let $\mathbf{x}_i^{sb,n}$ denote the best particle (i.e. the one associated with the lowest value of the objective function, in the case of minimization) among the neighbours of particle i. In Algorithm 5.1, the neighbourhood included all particles in the swarm, so that $\mathbf{x}_i^{sb,n} = \mathbf{x}^{sb}$ (independent of i). Such a neighbourhood is shown in the left panel of Fig. 5.3, where the discs represent the particles and the lines emanating from any disc determine the neighbours of the particle in question.

However, there are many alternatives to the fully connected neighbourhood. Another example is shown in the right panel of the figure, in which each particle

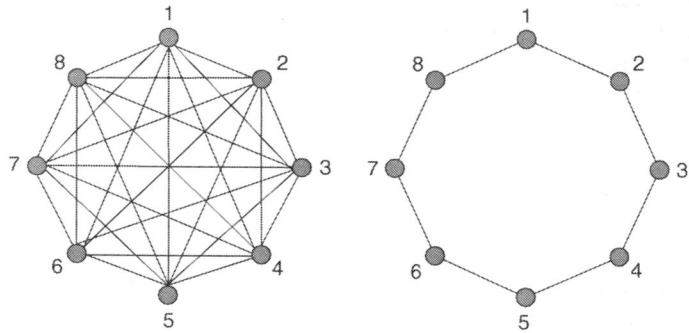

Figure 5.3: *Particle neighbourhoods in PSO, shown for the case $N = 8$. The left panel shows a fully connected neighbourhood and the right panel shows a neighbourhood with restricted connectivity.*

is only connected to its nearest neighbours on either side. Obviously, intermediate cases can be defined as well, in which each particle is connected to the K nearest neighbours on each side.

Note that, unlike the visibility spheres introduced in connection with boids (see Section 5.1.1), the topological constructs shown in Fig. 5.3 are defined in an abstract space different from the n-dimensional search space; two particles linked together as neighbours may be located at very different places in the search space. Furthermore, the neighbourhood structure normally remains fixed throughout optimization whereas, of course, the positions x_i in the *search* space vary with every iteration, according to eqns (5.14) and (5.15).

The definition of neighbourhood structures with restricted connectivity in PSO serves the same purpose as the various procedures introduced for the prevention of premature convergence in EAs, see Section 3.2.2. As a general rule, the introduction of neighbourhood structures with limited connectivity slows down convergence of the algorithm, but also generally allows the algorithm to cover a greater fraction of the search space, leading, in some cases, to improved results. Note that the different neighbourhoods overlap so that, if one particle displays particularly good performance, this result will eventually be communicated to the other neighbourhoods. For example, if particle 1 in the right panel of Fig. 5.3 finds a superior position, so that $x_1^{sb,n} = x_1$, it will attract particles 2 and 8 to that position and particle 2 will then communicate the result to the neighbourhood involving particles 2, 3 and 4, and, likewise, particle 8 will communicate the result to the neighbourhood consisting of particles 6, 7 and 8, etc.

5.3.3 Maintaining coherence

The choice of the parameters c_1 and c_2 strongly influences the trajectories of particles in the swarm. As noted above, a common choice is to take $c_1 = c_2 = 2$, in which case equal weight is given to the cognitive and social parts of the velocity equation, eqn (5.14). Other choices are possible as well. However, the sum of c_1 and c_2 should fulfil

$$c_1 + c_2 < 4. \tag{5.16}$$

One can prove (see Appendix B, Section B.4.1), after removing the stochastic components q and r in eqn (5.14), that the trajectories will remain bounded only if $c_1 + c_2 < 4$. However, even if $c_1 + c_2 < 4$, the trajectories of particles moving under the influence of Algorithm 5.1 will, in fact, diverge eventually, due to the influence of the random variables q and r. An example is shown in Fig. 5.4, which shows the position of a single particle integrated with $c_1 = c_2 = 1.5$. As is evident from the figure, even though $c_1 + c_2 < 4$, the trajectory eventually diverges.

Thus, in practice, the divergence of particle trajectories must somehow be controlled actively. The simplest way of doing so is to introduce a limit on particle

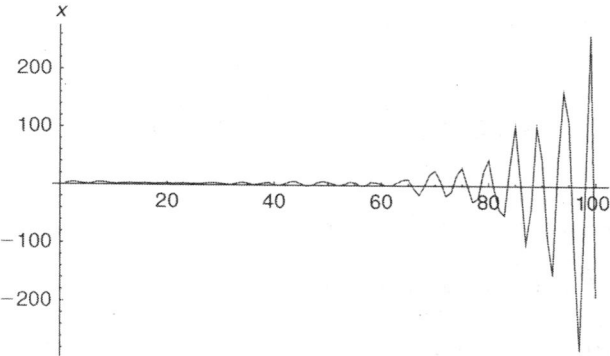

Figure 5.4: *Typical trajectory of a single particle, integrated (in one dimension) over 100 time steps, using eqns (5.14) and (5.15), with $c_1 = c_2 = 1.5$, $x^{pb} = 2$, $x^{sb} = 2.5$, $\Delta t = 1$. The integration was started at $x = 1$, with $v = 0$.*

velocities. Typically, the velocity of particle p_i is restricted such that

$$|v_{ij}| < v_{\max} = \frac{(x_{\max} - x_{\min})}{\Delta t}, \quad j = 1, \ldots, n. \tag{5.17}$$

Thus if, after an update using eqn (5.14), $v_{ij} > v_{\max}$, then v_{ij} is simply set equal to $v_{\max}$. Similarly if $v_{ij} < -v_{\max}$, v_{ij} is set equal to $-v_{\max}$. This, however, is not the only way to maintain swarm coherence. In fact, in the literature on PSO, there exist several studies [8, 9] concerning the use of **constriction coefficients** that modify the velocity equation (eqn (5.14)), the position equation (eqn (5.15)), or both. As an example, consider the modified rule

$$v_{ij} \leftarrow \chi \left(v_{ij} + c_1 q \left(\frac{x_{ij}^{pb} - x_{ij}}{\Delta t} \right) + c_2 r \left(\frac{x_j^{sb} - x_{ij}}{\Delta t} \right) \right), \quad j = 1, \ldots, n. \tag{5.18}$$

It can be shown [8] that the trajectories obtained using eqn (5.18) do not diverge if χ is taken as

$$\chi = \frac{2\kappa}{|2 - \xi - \sqrt{\xi^2 - 4\xi}|}, \tag{5.19}$$

where (note!) $\xi \equiv c_1 + c_2 > 4$ and κ is a parameter in the open interval $]0,1[$.

5.3.4 Inertia weight

As a further modification of Algorithm 5.1, one may introduce a parameter that determines the relative influence of previous velocities on the current velocity of

128 BIOLOGICALLY INSPIRED OPTIMIZATION METHODS

a particle. Consider the following modified velocity equation for particle p_i

$$v_{ij} \leftarrow w v_{ij} + c_1 q \left(\frac{x_{ij}^{pb} - x_{ij}}{\Delta t} \right) + c_2 r \left(\frac{x_j^{sb} - x_{ij}}{\Delta t} \right), j = 1, \ldots, n. \quad (5.20)$$

Here, w is referred to as the **inertia weight**. If $w > 1$, the particle favours exploration over exploitation, i.e. it assigns less significance to the cognitive and social components than if $w < 1$, in which case the particle is more attracted towards the current best positions. As in the case of EAs, exploration plays a more important role than exploitation in the early stages of optimization, whereas the opposite holds towards the end. Thus, a common strategy is to start with a value larger than 1 ($w = 1.4$, say), and then reduce w by a constant factor $\beta \in]0, 1[$ (typically very close to 1) in each iteration, until w reaches a lower bound (typically around 0.3–0.4). In fact, the use of an inertia weight is so common that we will define the standard PSO algorithm as Algorithm 5.1, but with the velocity equation (eqn (5.14)) replaced by eqn (5.20), and with a variation scheme for w of the kind just described.

Example 5.3
The experiments presented in Example 5.2 were repeated, using the same settings, but with an inertia weight varying from 1.4 to 0.4. In each iteration, w was reduced by a factor $\beta = 0.99$, until the lower limit of 0.4 was reached, after which w was kept constant at this level. The results are shown in Table 5.2. ∎

5.3.5 Craziness operator

The final PSO component that will be considered here is the so called **craziness operator**. This operator, which is typically applied with a given probability p_{cr}, sets the velocity of the particle p_i in question to a uniform random value within

Table 5.2: *The average and estimated standard deviations obtained from 100 samples of the distribution of the mean of the results generated using a standard PSO, applied to the two benchmark functions Ψ_2 and Ψ_4, for different values of c_1 and c_2. In each run, the inertia weight varied from 1.4 to 0.4; see Example 5.3. Note the striking difference in performance in the case of Ψ_4 compared to the results obtained in Example 5.2.*

Parameters		$\Psi_1(x_1,x_2)$		$\Psi_4(x_1,x_2,x_3,x_4)$	
c_1	c_2	Avg.	S.D.	Avg.	S.D.
2	2	3.000	0.000	1.036	0.304
1	2	3.000	0.000	0.494	0.233
2	1	3.000	0.000	0.231	0.109
1	1	3.000	0.000	1.045	0.307

the allowed range. Thus, if craziness is applied, the velocity of the particle changes according to

$$v_{ij} = -v_{\max} + 2rv_{\max}, \quad j = 1, \ldots, n, \tag{5.21}$$

where r is a random number in $[0, 1]$. The craziness operator, which in some way serves the same function as mutations in an EA, can be said to have a biological motivation: in flocks of birds one can observe that, from time to time, one bird suddenly shoots off in a seemingly random direction (only to re-enter the swarm shortly thereafter).

5.4 Discrete versions

In the PSO algorithms presented above, it is assumed that the variables x_j take values in a subset of $\mathbf{R}^n$. However, with only slight modifications, PSO can also be used in connection with integer programming problems, where the variables take integer values. Here, two discrete PSO algorithms will be considered. The first method considered, which is based on variable truncation, can be applied to general integer programming problems, whereas the second method (binary PSO) is applicable to problems in which the variables take binary values, i.e. $x_j \in \{0, 1\}, j = 1, \ldots, n$.

5.4.1 Variable truncation

The variable truncation PSO algorithm is straightforward: at initialization, random positions are generated as usual, but are then truncated (component by component) to their nearest integer values. The determination of new velocities and positions is then carried out exactly as in the continuous version. Once new positions have been obtained, each component of the position vector is truncated to the nearest integer value, and the objective function is computed. Moreover, since the internal workings of the algorithm are identical to the standard (continuous) PSO, the discussion above concerning specialized operators and parameter settings is valid in the discrete case as well. Let us now consider an example.

Example 5.4
Using PSO, find the minimum of the function

$$f(x_1, x_2) = \left(9x_1^2 + 2x_2 - 11\right)^2 + \left(3x_1 + 4x_2^2 - 7\right)^2, \tag{5.22}$$

with $x_1, x_2 \in \mathbf{Z}$. Set the search range for each variable to $[-100, 100]$.

Solution A variable truncation PSO algorithm was implemented as described in Section 5.4.1. The constants c_1 and c_2 were both set to 2, and the inertia weight varied, with $\beta = 0.99$, from an initial value of 1.4 to a final value of 0.4.

The PSO algorithm easily solved the problem, finding the minimum $\mathbf{x}^* = (1, 1)^T$, with $f(\mathbf{x}^*) = 0$. A total of 3,000 runs were carried out, using a swarm size of 20. The program stored the number of function evaluations needed in order to find the global minimum, an event that occurred in all runs. The number of function evaluations could be rather well approximated by a normal distribution, with a mean of 1,538 and a standard deviation of 282. ∎

5.4.2 Binary PSO

In certain cases of integer programming, it is necessary to restrict the particle positions to a given subset of $\mathbf{Z}^n$. One such case is binary programming, where the components x_j of the particle positions are restricted to $\{0, 1\}$. Note, as mentioned above, that while a restriction on velocities implies that positions will remain bounded, it does not restrict them to a range that can easily be pre-specified. Thus, a dedicated restriction mechanism for particle positions is needed. One such mechanism, introduced in Ref. [37] is shown in Algorithm 5.2. Here, the variables v_{ij} are updated as usual, but their interpretation is different: instead of directly modifying positions as in eqn (5.15), the v_{ij} are passed through an activation function σ (see Appendix A, Section A.2.1), given by

$$\sigma(v_{ij}) = \frac{1}{1 + e^{-v_{ij}}}, \qquad (5.23)$$

to generate the probability of setting x_{ij} equal to 1. With probability $1 - \sigma(v_{ij})$, x_{ij} is set to 0. Thus, with this modification, the positions are restricted to the desired set $\{0, 1\}$.

As in the case of the variable truncation PSO algorithm, the discussion concerning special operators and parameter settings in Section 5.3 is still mostly valid. One exception, however, is the truncation of velocities. In binary PSO, explicit truncation of velocities is used (rather than a constriction coefficient), such that the v_{ij} are typically restricted to the range $|v_{ij}| < v_{\max}$, where $v_{\max} \approx 4$, in order to avoid situations in which the probability of setting x_{ij} to any particular value (0 or 1) becomes too high. With $|v_{ij}|$ restricted to 4 or less, there is always a probability of at least $\sigma(v_{\max}) = 0.018$ of modifying the value of any position component x_{ij} for a particle p_i.

Some problems, such as decision-making problems involving a sequence of yes-or-no decisions, fall naturally into the category of binary programming, where binary PSO is directly applicable. Moreover, continuous optimization problems can also be solved using binary PSO, simply by employing a binary encoding scheme for the continuous variables, much as in a standard GA (see Chapter 3).

5.5 Applications

A few examples of PSO usage, focusing on applications involving neural networks, will now be given. A comprehensive list of PSO applications can be found in Ref. [58].

1. Initialize positions and velocities of the particles p_i:

   1.1 $x_{ij} \in \{0, 1\}$, (random), $i = 1, \ldots, N, j = 1, \ldots, n$
   1.2 $v_{ij} = -v_{\max} + 2rv_{\max}$, $i = 1, \ldots, N, j = 1, \ldots, n$

2. Evaluate each particle in the swarm, i.e. compute $f(\mathbf{x}_i)$, $i = 1, \ldots, N$.
3. Update the best position of each particle, and the global best position. Thus, for all particles p_i, $i = 1, \ldots, N$:

   3.1 if $f(\mathbf{x}_i) < f(\mathbf{x}_i^{pb})$ then $\mathbf{x}_i^{pb} \leftarrow \mathbf{x}_i$.
   3.2 if $f(\mathbf{x}_i) < f(\mathbf{x}^{sb})$ then $\mathbf{x}^{sb} \leftarrow \mathbf{x}_i$.

4. Update particle velocities and positions:

   4.1 $v_{ij} \leftarrow wv_{ij} + c_1 q \left(\frac{x_{ij}^{pb} - x_{ij}}{\Delta t} \right) + c_2 r \left(\frac{x_j^{sb} - x_{ij}}{\Delta t} \right)$, $i = 1, \ldots, N, j = 1, \ldots, n$
   4.2 Restrict velocities, such that $|v_{ij}| < v_{\max}(\approx 4)$.
   4.3 Compute $\sigma(v_{ij}) = \frac{1}{1 + e^{-v_{ij}}}$, $i = 1, \ldots, N, j = 1, \ldots, n$
   4.4 Generate a uniform random number $r \in [0, 1]$ and update x_{ij} as

   $$x_{ij} \leftarrow \begin{cases} 0 & \text{if } r > \sigma(v_{ij}) \\ 1 & \text{otherwise} \end{cases}$$

5. Return to step 2, unless the termination criterion has been reached.

Algorithm 5.2: *Binary particle swarm optimization. The main difference compared to Algorithm 5.1 is the interpretation of the velocities. As in Algorithm 5.1, it has been assumed that the best-ever position is used in step 3.2, and that the neighbourhood of any particle includes all other particles.*

5.5.1 Optimization of neural networks

Optimization of ANNs constitutes a very frequent application of PSO. Clearly, other stochastic optimization algorithms, for example EAs, can be applied to such problems as well and, furthermore, as shown in Appendix A, Section A.2.2, there are specialized methods, such as backpropagation, for optimizing ANNs. Thus, the use of PSO in connection with ANN applications may require some justification. Starting by a comparison with EAs, we note that such algorithms normally involve a crossover operator (see Chapter 3), which allows the algorithm to combine the genetic material of two individuals, and thus to explore widely separated parts of the search space. However, in the particular case of ANNs, such an operator is not very suitable. This is so since the computation in ANNs is of a distributed nature, i.e. it is carried out by a large number of basic computational elements (neurons, see Appendix A, Section A.2.1), each of which only performs a simple operation. Thus, in an ANN, the computation is intimately associated with the structure of the network. Now, if, during crossover, a chromosome encoding the network weights and biases (assuming that the number of neurons is fixed, for now) is split and joined with the parts of another chromosome, this operation is tantamount to cutting the ANN in two pieces and joining each piece with a part of another ANN. However, simply put, half an ANN does *not* carry out half of the computation of the complete network. Thus, joining partial ANNs by joining two chromosome parts from different individuals will most likely result in a network that does *worse* than the two original networks. An alternative would be to avoid crossover altogether by simply setting the crossover probability p_c to zero. Even though selection is done in proportion to fitness, the modifications of the networks would, in that case, be generated only by (random) mutations, and such an EA would normally run very slowly.

Turning to backpropagation (see Appendix A, Section A.2.2), possible disadvantages include its inability to cope with activation functions that are non-differentiable, and the possibility of the algorithm getting stuck at a local optimum in the search space (a risk that is reduced, however, if stochastic gradient descent is used rather than batch training, see Appendix A, Section A.2.2).

By their construction, PSO algorithms avoid the problems described above. First of all, the potential destructive effects of crossover are absent, since there is no equivalent to the crossover operator in (standard) PSO. However, rather than relying on random mutations, as an EA without crossover would have to do to a great extent, PSO, as we have seen above, introduces the notion of a velocity in the search space. While stochastic, the changes that *do* occur have a definite direction as determined by eqn (5.20), rather than being completely random. Furthermore, PSO algorithms do not easily get stuck in a local optimum and are also able to cope with non-differentiable activation functions, a property shared with EAs. Thus, it turns out that PSO algorithms are particularly well suited for optimizing ANNs. In several recent studies, the performance of PSO in ANN optimization has been compared with the performance of other algorithms. For example, in Ref. [26], a detailed comparison of PSO and backpropagation was carried out. The optimization algorithms were applied to the case of fitting a specific function, namely $g(x) = 2x^2 + 1$, using an FFNN with 2 input units, 4 hidden neurons and 1 output neuron.[3] Two cases were considered; in the first case, the training set was obtained by sampling the function at 21 equidistant points in the range $[-1, 1]$, and in the second case, 201 points were used.

In order to make the comparison between the two methods as fair as possible, the authors measured the number of elementary operations (additions and

---

[3] The standard logistic sigmoid was used as activation function for the neurons, and the output was rescaled to the interval $[-3, 3]$.

multiplications) executed during optimization. Thus, the ratio γ, given by

$$\gamma(E) = \frac{n_{\mathrm{bp}}(E)}{n_{\mathrm{pso}}(E)} \qquad (5.24)$$

was measured, where $n_{\mathrm{bp}}(E)$ and $n_{\mathrm{pso}}(E)$ denote the number of elementary operations needed to reach an error E for backpropagation and PSO, respectively.

Furthermore, in both methods, weight modifications were made based on the performance over the whole training set. In other words, batch training, rather than stochastic gradient descent, was used. As for the PSO, a fairly standard implementation was used with velocity modifications according to eqn (5.20) with $w = 0.8$ and $c_1 = c_2 = 2$. The number of particles was set to 25.

For the first case, with 21 data points, $\gamma(0.001)$ ranged from 1.40 to 6.12, indicating superior performance of the PSO relative to backpropagation. With 201 data points $\gamma(0.001)$ was reduced to 1.14. Thus, even in the latter case, the PSO outperformed backpropagation, albeit with a smaller margin.

5.5.1.1 Prediction of pollutant levels

In addition to pure comparisons of algorithms, as in Ref. [26], PSO has also been used in applied problems. An example, considered in Ref. [46], concerns the analysis of pollution in Hong Kong. As the authors point out, short-term prediction of pollutant levels in densely populated areas, such as Hong Kong, is of great importance, as it allows authorities to inform the public of the danger and also to take preventive action, for example in the form of traffic restrictions. Even though traffic restrictions may be associated with economic drawbacks, there can also be advantages, such as a reduction in the amount of hospital care needed to treat pollution-related medical problems.

In Ref. [46], a PSO was used for training networks with three layers of weights, on the problem of predicting the amount of respiratory suspended particles (RSPs) in the air, given seven input signals, namely the current concentrations of five substances (CO, NO, NO_2, NO_x and SO_2), the current concentration of RSPs and the current temperature. Thus, the FFNNs had seven inputs and one output. The number of neurons in both hidden layers was equal to eight. Two different prediction modes were tested: in the first case, the prediction concerned the level of RSPs in the next hour, with the input variables measured over the previous hour, whereas in the second case, the daily mean level of RSPs was predicted. In the latter case, the daily mean levels from the previous day were used as inputs. The available data, obtained from one of 14 monitoring stations around Hong Kong, were divided into a training set containing around 80% of the samples, and a test set with around 20% (see Appendix C, Section C.2). In the PSO, particle velocities were modified according to eqn (5.14), i.e. no inertia weight was used. The number of particles was equal to 30.

All runs lasted for a pre-specified number of iterations. Furthermore, a comparison with backpropagation was carried out. In order to make the comparison fair, the initial weights used by the backpropagation algorithm were taken as the weights obtained from the best particle in the first iteration of the PSO. The comparison was based on the number of evaluations of the objective function, taken as the mean square error of the predicted output compared to the desired output. As in Ref. [26], the PSO was found to outperform backpropagation, in the sense that the algorithm required fewer objective function evaluations than backpropagation in order to reach a given level of error, particularly in the second case (daily mean levels).

5.5.1.2 Prediction of elephant migration patterns

In Ref. [55], the problem of predicting elephant migration was considered. During foraging, large herbivores, such as elephants, consume very large quantities of plants, and the impact on the environment can be significant, particularly if the animals are confined to a rather small region, for example a game reserve of small size. Surveying the state of the vegetation in a game reserve is a time-consuming task. In addition, the problem of determining which areas to prioritize for survey is a dynamic one, since the habitat selection of elephant herds is of course dependent on the amount of vegetation available. Thus, for conservation management, it is very important to be able to predict the movement of elephant herds.

In Ref. [55], the authors considered the movement of elephants in the Pongola Game Reserve in South Africa. At the time of the study (2000–2002) this reserve, with an area of around 74 km^2 contained 45 elephants, of which 37 belonged to one family group. A female member of this group was fitted with a GPS, allowing the researchers to measure the movement of the family group over time. Once the data had been collected, ANNs were trained to predict the position of the herd on a daily basis. It was found that the best results were obtained using two ANNs, with a single output each, for predicting the position components ($x \equiv x_1$ and $y \equiv x_2$, respectively) of the herd. In this study, recurrent neural networks of the Elman kind were used (see Appendix A, Section A.2.3), with feedback connections from the hidden neurons to context units in the input layer. As mentioned in Appendix A, the recurrent connections provide the ANN with a dynamic memory, such that the output of the nextwork depends not only on the current inputs but also on inputs previously presented to the network. The structure of the ANNs used in Ref. [55] are shown schematically in Fig. 5.5. As can be seen from the figure, the two networks predicting[4] $x_i(t+1), i = 1, 2$ consisted of 20 input units, four of which measured the four previous positions $x_i(t)$, $x_i(t-1)$, $x_i(t-2)$, and $x_i(t-3)$ and the remaining 16 conveyed the previous output from the hidden layer, the output of which was fed to the single output neuron.

The data set consisted of measurements of x_1 and x_2 for 360 days. The first $N_{tr} = 180$ data points were used during training, and the remaining points were used for testing. The training was carried out using a standard PSO, with velocity updates according to eqn (5.20), with $c_1 = 1.5$ and c_2 increasing linearly from 0.1 to 1.5 over the 350 iterations carried out during training. Thus, in early stages of training, the social component c_2 had little impact, preventing premature convergence towards

---

[4] The predicted value of a variable x_i is here denoted $\hat{x}_i$.

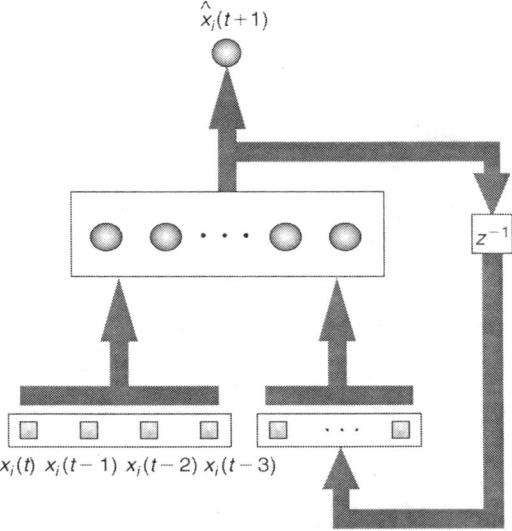

Figure 5.5: *The ANNs used in Ref. [55] to predict elephant movements. Two networks were used, one for each position component (x_1 and x_2) for the female fitted with a GPS. The z^{-1} element in the figure indicates that the signals entering the 16 context units in the input layer correspond to the signals from the 16 hidden-layer neurons in the previous time step. The network weights are not explicitly shown.*

the (current) global best solution. The inertia term was kept fixed at 0.8 for the first 75% of the training epochs. It was then reduced linearly to 0.4 over the remaining iterations. The objective function for network $i, i = 1, 2$ was taken as the mean square error in the predicted positions

$$F_i = \frac{1}{N_{\text{tr}}} \sum_t (\hat{x}_i(t) - x_i(t))^2. \tag{5.25}$$

After training the networks, using 40 particles for the PSO, the root mean square error ($\sqrt{F_i}$), which can be taken as a measure of the average deviation in position, was found to be 0.93 and 2.26 km for x_1 and x_2, respectively, over the training set. For the test set, the corresponding numbers were 0.82 and 1.86 km. However, as the authors noted, since the current output of an RNN depends on previous inputs (see Appendix A, Section A.2.3), prediction errors may accumulate in the network. In order to prevent such problems, a predictor–corrector technique was used, taking as output $\hat{x}_i^*(t)$, defined as

$$\hat{x}_i^*(t) = \hat{x}_i(t) + \frac{1}{2}(x_i(t) - \hat{x}_i(t)). \tag{5.26}$$

The use of the predictor–corrector method dramatically improved the prediction results, reducing the error over the test set to 0.58 and 0.91 km for x_1 and x_2, respectively.

5.5.2 Optimization of cancer chemotherapy

The problem of administering drugs at an optimal rate during cancer chemotherapy was studied in Ref. [57]. The problem is difficult, since the optimal doses are near the levels at which side effects become unacceptable; drugs used in chemotherapy are highly toxic and, since they are delivered through the blood stream, they affect not only the tumour that is their target, but also other tissues in the body. Hence, it is crucial to control the drug doses carefully. In chemotherapy, drugs are usually administered through a discrete dosage program, such that n doses of m substances are given at times $t_i, i = 1, \ldots, n$. The concentration at time i of drug j is denoted c_{ij}. The dosage program thus has a total of $n \times m$ tunable parameters.

The variation in tumour size $N = N(t)$ can be modelled using the Gompertz growth model [23], augmented with linear cell loss (as a result of the treatment)

$$\frac{dN}{dt} = N \left(\lambda \ln \frac{a}{N} - \sum_{j=1}^{m} \kappa_j \sum_{i=1}^{n} c_{ij} \left(\Theta(t - t_i) - \Theta(t - t_{i+1}) \right) \right), \qquad (5.27)$$

where N denotes the number of tumour cells, λ and a are the parameters of tumour growth, and $\Theta(t)$ is Heaviside's step function, taking the value 0 if $t < 0$, and 1 otherwise. Thus, the final term on the right-hand side indicates that the drug concentration c_{ij} is considered to be constant between $t = t_i$ and $t = t_{i+1}$. t_{n+1} can be defined as $2t_n - t_{n-1}$. Now, the toxicity of the drugs imply certain constraints on their delivery. In Ref. [57], four constraints were defined. Let C denote the matrix with elements c_{ij} (see above). First of all, there are limits on the maximum instantaneous dose, leading to the constraints $g_1(C)$ defined as

$$g_1(C) = \left\{ c_{ij} - c_j^{\max} \leq 0, \quad i = 1, \ldots, n, \; j = 1, \ldots, m \right\}. \qquad (5.28)$$

Second, there is also a maximum cumulative dose c_j^{mcd} for each drug, giving the constraints

$$g_2(C) = \left\{ \sum_{i=1}^{n} c_{ij} - c_j^{\mathrm{mcd}} \leq 0, \quad j = 1, \ldots, m \right\}. \qquad (5.29)$$

The maximum permissible tumour size $N^{\max}$ provides the third set of constraints

$$g_3(C) = \{ N(t_i) - N^{\max} \leq 0, \quad i = 1, \ldots, m \}. \qquad (5.30)$$

Finally, the constraints on the toxic side effects of the chemotherapy are expressed as

$$g_4(C) = \left\{ \sum_{j=1}^{m} \eta_{kj} c_{ij} - c_k^{tse} \leq 0, \quad i = 1, \ldots, n, \quad k = 1, \ldots, K \right\}, \quad (5.31)$$

where η_{kj} measures the damaging effect of drug j on tissue or organ k. Obviously, the aim of chemotherapy is to reduce the size of the tumour as much as possible. Thus, the optimization problem can be formulated as

$$\text{minimize } F(C) \equiv \sum_{i=1}^{n} N(t_i), \quad (5.32)$$

subject to the constraints $g_1 - g_4$. This optimization problem was attacked, in Ref. [57], using both a GA with binary encoding scheme and (several versions of) PSO. The authors compared both the ability of each algorithm to find a feasible solution, and the quality of the solutions found. The population size for the GA was set to 50, as was the number of particles for the PSO. The GA employed roulette-wheel selection and elitism, in which two copies of the best individual in any given generation were transferred unchanged to the next generation. Two-point crossover was used. For the PSO, velocities were updated using eqn (5.20), with a fixed inertia weight. Each run consisted of 25,000 evaluations of the objective function. Averaged over 30 runs with each algorithm, it was found that the PSO discovered feasible solutions much faster than the GA: in fact, in some runs, the GA did not find any feasible solution at all, whereas the PSO typically found such solutions after rather few iterations, and never failed. As for the quality of the solutions, the PSO again showed superior performance, but with a smaller margin. A possible explanation of the difference can be that the best solutions might be located near boundaries of feasible regions, which the PSO can handle quite well, but where the GA, employing crossover, often generates new individuals outside the feasible regions.

Exercises

5.1 Write a computer program implementing the boids model, described in Section 5.1.1 in two dimensions, and rerun the experiments described in Example 5.1. Try to find some parameter values that lead to a bird-like swarming behaviour over long periods of time.

5.2 Implement the basic PSO described in Algorithm 5.1 and apply it to the problem of finding the minimum of the function

$$f(x_1, x_2) = 1 + \left(-13 + x_1 - x_2^3 + 5x_2^2 - 2x_2\right)^2$$
$$+ \left(-29 + x_1 + x_2^3 + x_2^2 - 14x_2\right)^2, \quad (5.33)$$

in the (initial) search range $x_i \in [-10, 10]$, $i = 1, 2$.

5.3 If the number of dimensions is small, i.e. three or less, a swarm can be visualized. Repeat the previous problem, plotting the motion of the entire swarm during the run. Also, carry out the same analysis for a PSO applied to the minimization of the function $\Psi_2^{[3]}(x_1, x_2, x_3)$ in Appendix D.

5.4 Consider the results for Ψ_4 from Example 5.2, shown on the right side of Table 5.1. Can the null hypothesis $\mu_i = \mu_j$ (see Appendix C) be rejected for any pair (i, j) of parameter sets? Assume normal distributions with unequal variance.

5.5 Plot the trajectory of a single particle, following the simplified deterministic PSO defined in eqns (B79) and (B80), for various values of c_1 and c_2 in the range [0, 3], and verify (numerically) that the trajectory remains bounded only if $c_1 + c_2 < 4$.

5.6 Implement integer PSO using variable truncation, as described in Section 5.4.1, and apply the program to find the minimum of the function [60]

$$f(x_1, x_2) = 2x_1^2 + 3x_2^2 + 4x_1 x_2 - 6x_1 - 3x_2, \tag{5.34}$$

with $x_i \in \mathbb{Z}$, $i = 1, 2$. Set the initial search range to $[-100, 100]$ for each variable.

Chapter 6

Performance comparison

In this brief concluding chapter, we shall compare the various stochastic optimization methods considered in earlier chapters. Clearly, there is no limit on the number of benchmark problems that could be considered in such a comparison, and neither is there any limit on the number of possible configurations (components and parameter sets) of the tested algorithms. Thus, trying to cover all problems and configurations of algorithms, or even a large subset thereof, would be a futile exercise. Furthermore, the NFL theorem mentioned in Section 1.2 implies that, averaging over all problems, no algorithm would actually outperform any other. Yet, in practice it is found, for typical optimization problems, that the stochastic optimization methods covered in this book easily outperform, for example, a random search (an illustration is given below), and also that some parameter settings work better than others or, rather, that there are certain parameter settings that should really be avoided.

Thus, the aim of this chapter is to give the reader an indication of the suitability of different algorithms and parameter settings for a few classes of problems. In doing so, we continue the analysis presented, for instance, in Examples 3.5 and 5.3, but this time with primary emphasis on the comparison of different algorithms rather than the parameter settings of the individual algorithms that, to a great extent, will be based on the findings in Chapters 3–5. However, not all algorithms will be applied to all problems; even though it would be possible, say, to apply ACO to the optimization of a mathematical function $f(x_1, \ldots, x_n)$ by means of a chain construction graph (see Chapter 4), it is more natural to apply either a GA or PSO to this type of problem. Likewise, PSO could be adapted for solving the TSP but, in this case, a comparison of ACO and GAs is more relevant, as these algorithms are directly applicable to the problem.

Analyses of this type occur frequently in the literature, particularly in connection with the many hybrid algorithms that are often introduced in attempts to combine the positive features of the individual algorithms. One example, among many, of such methods is the **breeding swarms** algorithm introduced in Ref. [69], which combines GAs and PSO. The comparisons presented here, however, will not include hybrid algorithms.

6.1 Unconstrained function optimization

The problem of unconstrained function optimization will be considered using GAs and PSO. Even though there are no explicit constraints in this case, the search is typically limited to some range $x_i \in [x_{\min}, x_{\max}]$, $i = 1, \ldots, n$, see, for example, the description of benchmark functions in Appendix D. A GA does, by its construction, introduce limits on the allowed range of the variables x_i, $i = 1, \ldots, n$, and it is therefore well suited for this type of range constraints. By contrast, in (standard) PSO, only the velocities are strictly limited, but the positions may venture outside their original range, as noted in Chapter 5. Thus, even when care is taken to make the conditions as equal as possible for the two algorithms being compared, there will always be some subtle differences regarding, for example, which regions of the search space will actually be covered.

In addition, a comparison with random search (RS) will be carried out. This algorithm simply generates random vectors $\mathbf{x}$, with $x_i \in [x_{\min}, x_{\max}]$, and computes the value of the objective function $f(\mathbf{x})$ for each new vector, keeping track of the best result found.

Two benchmark functions were considered, namely $\Psi_2^{[5]}$ and $\Psi_5^{[10]}$ (see Appendix D). The two parameters in $\Psi_5^{[10]}$ were set to $a = 0.05, b = 10$. For the GA, tournament selection was used throughout, with tournament size equal to two and $p_{\text{tour}} = 0.90$. The crossover probability was also set to 0.90. For the runs using binary encoding, the mutation rate was set to $1/m$, where m as usual denotes the chromosome length. Thirty bits per variable were used, so that p_{mut} thus equalled $1/30n$. In cases with real-number encoding, p_{mut} was equal to $1/n$. In those simulations, real-number creep was used: with probability p_{cr} a creep mutation with range C_r (see eqn 3.24) was carried out instead of an ordinary full-range mutation. The latter kind of mutation was thus carried out with probability $1 - p_{\text{cr}}$.

The PSO was of standard type, with $c_1 = c_2 = 2$ in all runs. The inertia weight (w) varied from 1.4 to 0.4, with $\beta = 0.99$. $\mathbf{x}^{\text{sb}}$ was defined as *best so far*, and the neighbourhood topology was fully connected, as in the left panel of Fig. 5.3. For all parameter settings, regardless of the optimization method used, 10,000 function evaluations were carried out per run.

The results are summarized in Table 6.1, which shows averages and estimated standard deviations of the best results found over 1,000 runs. The PSO showed best performance for both problems (note that the second problem involved *maximization*). This does not imply that PSO would always do better than a GA on the task of unconstrained function optimization: only a few parameter settings were tested, and it is likely that there exist some parameter combination for which the GA would achieve better results than in the table (though not necessarily better results than

Table 6.1: *Results obtained for unconstrained optimization of the functions $\Psi_2^{[5]}$ (minimization) and $\Psi_5^{[10]}$ (maximization).*

$\Psi_2^{[5]}(x_1,\ldots,x_5)$					
Method	Settings	N	I	Avg.	S.D.
GA	Binary	40	250	19.06	40.64
GA	Binary	100	100	3.849	7.138
GA	Binary	250	40	2.401	3.403
GA	RN, $p_{cr}=0.8$, $C_r=0.2$	250	40	5.404	1.682
GA	RN, $p_{cr}=0.8$, $C_r=0.02$	250	40	1.829	1.244
PSO	–	20	500	0.727	1.161
PSO	–	40	250	0.689	0.845
PSO	–	100	100	2.600	1.325
PSO	–	250	40	55.90	31.64
RS	–	1	10,000	95.19	48.74

$\Psi_5^{[10]}(x_1,\ldots,x_{10})$					
Method	Settings	N	I	Avg.	S.D.
GA	Binary	40	250	0.817	0.058
GA	Binary	100	100	0.875	0.048
GA	Binary	250	40	0.875	0.038
GA	RN, $p_{cr}=0.8$, $C_r=0.2$	250	40	0.747	0.034
GA	RN, $p_{cr}=0.8$, $C_r=0.02$	250	40	0.851	0.044
PSO	–	20	500	0.898	0.049
PSO	–	40	250	0.881	0.049
PSO	–	100	100	0.786	0.035
PSO	–	250	40	0.684	0.030
RS	–	1	10,000	0.687	0.026

I *denotes the number of iterations.* RN = *real-number encoding.* p_{cr} *denotes the creep probability and* C_r *the creep rate.* RS = *random search. See the main text for a complete description of the settings used for the different methods.*

PSO). Also, what amounts to a *standard* algorithm is, of course, largely a matter of definition. In this book, the standard PSO includes variation in the inertia w, and it therefore tackles the tradeoff between exploration and exploitation in a more sophisticated way than a standard GA, something that may also explain the superior results obtained using PSO.

However, some tentative conclusions can, perhaps, be drawn from Table 6.1. First of all, it is interesting to note that the GA does better with large population

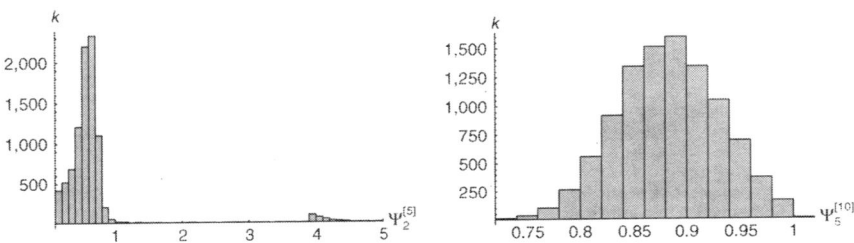

Figure 6.1: *The left panel shows the distribution of results obtained by running a PSO 10,000 times, with the aim of minimizing $\Psi_2^{[5]}$, with $N=40$ and $I=250$. The right panel shows the distribution of results obtained for $\Psi_5^{[10]}$ using the same settings. In the left panel, the bin width in the histogram was 0.1, whereas a bin width of 0.02 was used in the right panel. k denotes the number of runs per bin.*

size, whereas, as noted also in Chapter 5, the PSO generally achieves its best results in runs involving small swarms. The tendency of the crossover operator to cause inbreeding may explain why the GA achieves better results for larger population sizes. Of course, in PSO, there is no crossover operator, and therefore no risk of inbreeding.

For the GA with real-number encoding, the best results were obtained using a rather small creep rate C_r. Comparing binary and real-number encoding, no particular conclusions can be drawn, though: real-number encoding achieves better results than binary encoding for $\Psi_2^{[5]}$, but the opposite holds for $\Psi_5^{[10]}$.

In both problems, the biologically inspired methods, i.e. the GA and the PSO, easily outperformed random search. This does not, in any way, invalidate the NFL theorem which, of course, is a statement regarding optimization of *all* possible benchmark functions. Yet, the result does tell us that, in problems encountered in practice, it is worthwhile to apply, for example, one of the biologically inspired optimization methods discussed in this book. Note, however, that the parameter settings *do* matter. In fact, even though the PSO gave the best result in both of the optimization problems considered, it also gave the *worst* result (in the second problem), even worse than random search!

As a final point of observation, we note that the distribution of results differs quite strongly from a normal distribution in some cases, as illustrated in the left panel of Fig. 6.1. In those cases, the measured standard deviation typically exceeds the average. However, for more difficult problems, the distribution typically resembles a normal distribution quite well, as illustrated in the right panel of Fig. 6.1. One must be careful before applying hypothesis testing as described in Appendix C since, in such tests, it is assumed that the variables (in this case the fitness values) follow normal distributions. In cases where the distribution differs significantly from a normal distribution, one might consider studying the sampling distribution of the *mean*, also described in Appendix C.

6.2 Constrained function optimization

As mentioned in Chapter 2, many methods have been developed for solving constrained optimization problems within the framework of classical optimization. Furthermore, methods for solving such optimization problems have also been developed in the field of stochastic optimization, see, for example, Ref. [49]. Here, we shall apply the penalty method, described in Section 2.4.2.

Two different problems will be considered. In the first problem, the objective is to minimize

$$f_1(x_1, x_2) = (x_1 - 2)^2 + (x_2 - 1)^2, \tag{6.1}$$

subject to the constraints

$$g_{11}(x_1, x_2) = x_1 + x_2 - 2 \leq 0, \tag{6.2}$$

and

$$g_{12}(x_1, x_2) = x_1^2 - x_2 \leq 0. \tag{6.3}$$

The exact global minimum for this problem occurs at $(x_1^*, x_2^*)^T = (1, 1)^T$, where the function takes the value $f_1(1, 1) = 1$. The second problem consists of minimizing

$$f_2(x_1, x_2) = (x_1 - 10)^3 + (x_2 - 20)^3, \tag{6.4}$$

subject to the constraints

$$g_{21}(x_1, x_2) = 100 - (x_1 - 5)^2 - (x_2 - 5)^2 \leq 0, \tag{6.5}$$

and

$$g_{22}(x_1, x_2) = -82.81 + (x_1 - 6)^2 + (x_2 - 5)^2 \leq 0. \tag{6.6}$$

In this problem, the variables are constrained to the intervals $x_1 \in [13, 100]$ and $x_2 \in [0, 100]$. Now, in the case of a GA, the range constraints can easily be taken into account by specifying exactly the required range when decoding the chromosomes. However, for the (standard) PSO, only the velocities are strictly limited, not the positions. Thus, in order to make sure that the algorithm does not attempt to find an optimum outside the allowed intervals, the range constraints were represented explicitly by the four constraint functions

$$g_{23}(x_1) = 13 - x_1 \leq 0, \tag{6.7}$$

$$g_{24}(x_1) = x_1 - 100 \leq 0, \tag{6.8}$$

$$g_{25}(x_2) = -x_2 \leq 0, \tag{6.9}$$

Table 6.2: *Results obtained for the first constrained optimization problem, involving the minimization of $f_1(x_1,x_2)$ subject to two constraints.*

Method	Settings	N	I	$r(\mu=100)$	$r(\mu=1,000)$
GA	Binary	100	200	0.64	0.35
GA	Binary	250	80	0.95	0.75
GA	RN, $p_{cr}=0.8$, $C_r=0.10$	250	80	0.76	0.57
PSO	–	40	500	1.00	1.00
PSO	–	100	200	1.00	1.00

The fraction r of successful runs for each algorithm and parameter setting, averaged over 100 runs, is shown, for two different values of μ. In each run, 20,000 function evaluations were carried out. See Table 6.1 and Section 6.1 for a description of the parameter settings.

and

$$g_{26}(x_2) = x_2 - 100 \leq 0. \tag{6.10}$$

The global minimum for this problem occurs at $(x_1^*,x_2^*)^T = (14.095, 0.843)^T$, where the function takes the value $-6,961.813$.

Once the constraints have been defined, stochastic optimization methods can be applied as usual, but with $f_p(\mathbf{x};\mu)$, rather than $f(\mathbf{x})$, as the objective function, see eqn (2.67). However, the presentation of results is less straightforward, as one must take into account both the value of the objective function and the penalty term.

The problems were attacked using both GAs and PSO. For the first problem described above, which turned out to be quite simple, the results are presented in Table 6.2. The table shows the success rate (measured over 100 runs for each parameter setting), defined as the fraction of runs in which (1) the best (smallest) value f_1^{min} of the objective function fulfilled $f_1^{min} < 1.005$ and (2) the penalty term, including the factor μ as in eqn (2.66), fulfilled $p < 0.003$. In all runs, 20,000 function evaluations were carried out. For the PSO all runs were successful, according to these criteria. Yet, the table belies the fact that for $\mu = 100$ the PSO often got stuck at a some, albeit small, distance from the optimum whereas, for example, the GA with real-number encoding found more or less the exact optimum in those runs in which it was successful. However, for $\mu = 1,000$, the PSO really did better, by all criteria, reaching an average of less than 1.0003 and with an average penalty of around 2.2×10^{-4}.

The results obtained for the second problem are given in Table 6.3. The table shows the average $\bar{f}_2$, over 100 runs, of the best values found for the objective function, as well as the average penalty $\bar{p}$, according to eqn (2.66) for the best individual, or particle, in each run. Interestingly, the GA using binary encoding failed to produce a meaningful average since, in some (rare) runs, it failed altogether, only reaching (positive) values of the objective function several powers of ten above the best values. Thus, the results from those runs have been omitted from the table,

Table 6.3: *Results obtained for the second constrained optimization problem, involving the minimization of $f_2(x_1, x_2)$ subject to six constraints.*

Method	Settings	N	I	$\overline{f}_2, \overline{p}\,(\mu=10^5)$	$\overline{f}_2, \overline{p}\,(\mu=10^6)$
GA	RN	100	200	−6,968.73, 37.31	−6,735.30, 8.09
GA	RN	250	80	−6,598.83, 65.63	−6,671.18, 4.73
PSO	−	40	500	−6,968.59, 6.76	−6,962.57, 0.68
PSO	−	100	200	−6,968.47, 6.52	−6,959.69, 0.56

The table shows the average $\overline{f}_2$ over 100 runs of the best values found, together with the average $\overline{p}$ of the penalty term. In each run, 20,000 function evaluations were carried out. See Table 6.1 and Section 6.1 for a description of the parameter settings. For the GA, p_{cr} was equal to 0.8 and C_r was equal to 0.02.

even though the successful runs obtained with binary encoding reached roughly the same results as those obtained using real-number encoding.

Note that some of the results for the average $\overline{f}_2$ are actually below the global minimum and are consequently associated with a rather large penalty term. The best results were obtained using the PSO with $\mu = 10^6$. In this case, near-optimal results were obtained, with very small penalties.

6.3 Optimization of feedforward neural networks

Optimization of FFNNs is really a special case of unconstrained function optimization, since the computation carried out by such a network can be described in the form of a mathematical function, in which the network weights constitute the variables. However, it is an important special case that merits special consideration, particularly since there exist specialized methods, for example backpropagation, for the optimization of FFNNs.

Here, we shall compare three methods, namely GAs, PSO and backpropagation, applied to the problem of making an FFNN represent the three–parity function, described in Example A.2. For backpropagation, stochastic gradient descent (see Appendix A) was applied, and the algorithm was allowed to run for 20,000 epochs. The initial weight range was set to $[−0.20, 0.20]$.

A standard GA was used, with 16 variables encoding the weights and bias terms of the 3–3–1 networks considered. In runs with real-number encoding, p_{mut} was set to $1/n = 1/16$. The probability of creep mutations was set to 0.8, and the creep rate to 0.005. In runs with binary encoding, the number of bits per variable was equal to 30 and the mutation rate was set as $p_{mut} = 1/m = 1/480$. Tournament selection was used in all GA runs, with a tournament size of two and with $p_{tour} = 0.90$. The range of the variables was set to $[−10, 10]$.

The PSO algorithm was also a standard one, with the inertia weight w dropping from an initial value of 1.4 to a final value of 0.4, with $\beta = 0.995$. In all runs, the

Table 6.4: *Comparison of results obtained for the 3–parity problem. The fraction of successful runs* (r) *as well as the average and estimated standard deviation of the RMS error* E_{rms} *are shown.*

Method	Settings	N	I	r	Avg.	S.D.
BP	$\eta = 0.15$	1	20,000	0.87	0.0108	0.0012
BP	$\eta = 0.30$	1	20,000	0.93	0.0068	0.0004
GA	RN	100	200	1.00	0.0008	0.0004
GA	Binary	100	200	0.67	0.0042	0.0004
PSO	–	40	500	0.90	0.0002	0.0012
PSO		100	200	0.96	0.0000	0.0000

BP = *backpropagation.* I *denotes the number of iterations (i.e. the number of generations, in the case of the GA).*

parameters c_1 and c_2 were both set to 2. The initial range of the variables was set to $[-10, 10]$.

The objective function for the GA and the PSO algorithms was taken as the RMS error, E_{rms}, over the training set, consisting of eight elements, as shown in Table A.1. In all cases, the activation function of the neurons was the standard sigmoid, defined in eqn (A6), with $c = 2$.

Twenty-thousand evaluations were used, both for the GA and the PSO algorithms. The comparison is thus slightly unfair, in favour of backpropagation, since this algorithm must carry out the backward pass through the network, which requires more operations than forming new individuals in the GAs or new particle positions for PSO.

The results are summarized in Table 6.4. Forming a simple average of the errors obtained would not be meaningful as some runs were utter failures, in which the error got stuck at high levels (0.25 or more). Instead, the table lists the fraction of successful runs for each method, defined as runs in which the error E_{rms} dropped to 0.02 or less. For *those* runs, the average and estimated standard deviation are also listed. In total, 30 runs were carried out for each method and parameter setting. Note the near-perfect results obtained with PSO.

6.4 The travelling salesman problem

As a final example, we shall return to the TSP, which served as an example in the description of ACO in Chapter 4. Here, two methods will be compared, namely ACO and GAs. The application of ACO to the TSP is straightforward, and follows the description given in Chapter 4. Applying a GA to the TSP is less straightforward, perhaps, and we shall therefore begin with a brief explanation. The main issue is the encoding scheme that, in this case, must describe paths rather than mathematical variables. This can be achieved by using the permutation encoding described in Section. 3.2.1, in which the chromosome of an individual simply consists of a permutation (i.e. an ordering) of the node (city) indices $1, \ldots, n$. For example, in a simple case involving $n = 7$ nodes, a possible chromosome would be $c = (7, 5, 4, 3, 6, 2, 1)$. This chromosome would encode a tour starting in node v_7, then going to node v_5 etc. When node v_1 has been reached, the tour concludes with a return to node v_7. This is simple enough and, to a great extent, the standard operators of a GA can be used as in any other application. Thus, the flow of the algorithm essentially follows the description given in Algorithm 3.2. However, the modification of chromosomes taking place during crossover and mutation must (for the TSP) map valid paths onto other valid paths, and therefore one must use customized versions of these two operators. The standard mutation operator can be replaced by **swap mutation**. Here, if a gene mutates, an event that, as usual, occurs with probability p_{mut}, a second gene is chosen randomly along the chromosome, and the values of the two genes (i.e. the node indices) are swapped.

Crossover is a bit more complicated, and several different operators suitable for the TSP have been proposed. One such operator is **order crossover**, in which substrings are exchanged between the two parents while keeping the order among those nodes which are not part of the exchanged substrings. For example, consider the two parent chromosomes (7, 1, 2, 4, 6, 3, 5) and (2, 3, 5, 1, 4, 7, 6). Two crossover points are chosen randomly, for instance between genes three and four and between genes six and seven. The corresponding substrings are 4, 6, 3 and 1, 4, 7. These are inserted in the (initially empty) offspring chromosomes which then take the form (-, -, -, 1, 4, 7, -) and (-, -, -, 4, 6, 3, -), where the - indicate an as yet empty slot. The first offspring will then have its empty slots filled in using the remaining string of the first parent. Thus, the nodes 1, 4, and 7 are removed from the first string resulting in a substring (2, 6, 3, 5). The elements of this substring are then inserted, in order, in the chromosome of the first offspring, starting from the first empty slot after the substring (1, 4, 7). The resulting chromosome is (6, 3, 5, 1, 4, 7, 2). Similarly, the second offspring takes the form (5, 1, 7, 4, 6, 3, 2).

The two methods were applied to an instance of the TSP with $n = 100$ nodes, the coordinates $(x_{1i}, x_{2i})^{\text{T}}$ of which were generated according to

$$x_{1i} = 0.1((9 + 13i^2) \bmod 200), \qquad (6.11)$$

and

$$x_{2i} = 0.1((7 + 1327i) \bmod 200). \qquad (6.12)$$

The location of the nodes is illustrated in Fig. 6.2, together with the shortest path found, with length $D^* = 168.60$. It turns out that this instance of TSP is, in fact, quite challenging. Table 6.5 summarizes the results obtained with different algorithms and parameter settings. For AS, the initial pheromone level τ_0 was set as N/D^{nn}, where D^{nn}, the length of the path obtained by moving, at each step, to the nearest neighbour, was around 220 length units. For MMAS, τ_{max} and τ_{min} were determined as described in Section 4.2.2, and the initial pheromone level was set to

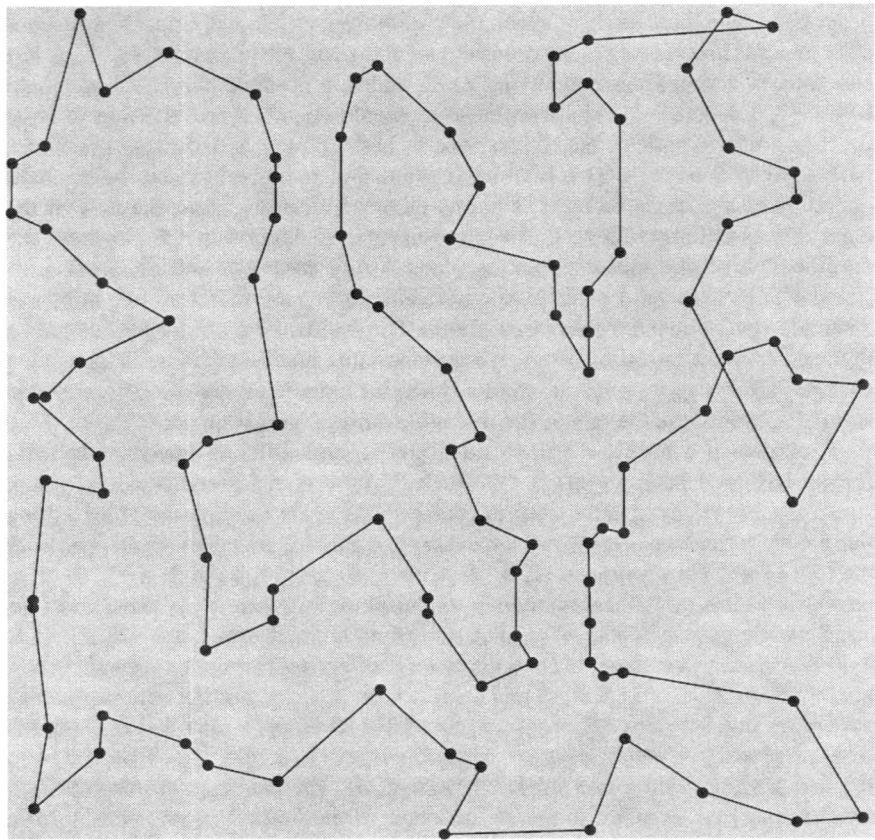

Figure 6.2: *An instance of the TSP, with 100 nodes, used in the performance comparison involving ACO and GAs. The length of the path shown, which is the shortest path discovered for this particular instance of the TSP, is $D^* = 168.60$ length units.*

$\tau_{\max}$. The parameter ρ was set to 0.5 both for AS and MMAS. For the GAs, tournament selection was used, with a tournament size of two and with $p_{\text{tour}} = 0.90$. The probabilities for (order) crossover and (swap) mutation are given in the table. Note that crossover does not improve the performance in this case. GA-R refers to a GA in which the initial paths were generated completely randomly. As can be seen in the table, this algorithm did not do very well. However, comparing it directly with ACO is quite unfair since ACO is very likely to find the nearest-neighbour path almost immediately. This is so, since, when all pheromone levels are equal, the generated path is determined solely by the factors $\eta_{ij}^{\beta} = (1/d_{ij})^{\beta}$, see eqn (4.3). Thus, for any given step, the most likely target is the nearest city and, particularly for large values of β, the algorithm is virtually guaranteed to find the nearest-neighbour path, which is much shorter than a randomly generated path.

Table 6.5: *Results obtained by applying different algorithms to the instance of the TSP shown in Fig. 6.2.*

Method	Settings	N	I	Avg.	S.D.
AS	$\alpha=1, \beta=2, \tau_0=0.454$	100	100	179.68	2.81
AS	$\alpha=1, \beta=5, \tau_0=0.454$	100	100	173.84	1.89
MMAS	$\alpha=1, \beta=5, \tau_0=0.00909$	100	100	193.58	2.67
GA-R	$p_c=0.00, p_{mut}=0.01$	50	200	577.18	16.94
GA-R	$p_c=0.00, p_{mut}=0.03$	50	200	658.40	25.45
GA-R	$p_c=0.20, p_{mut}=0.01$	100	100	647.57	18.73
GA-NN	$p_c=0.00, p_{mut}=0.01$	50	200	217.89	16.04
GA-NN	$p_c=0.00, p_{mut}=0.01$	100	100	211.67	12.62
GA-NN	$p_c=0.20, p_{mut}=0.01$	100	100	216.68	10.29
GA-NN	$p_c=0.00, p_{mut}=0.01$	200	50	206.57	10.13

For each method and parameter setting, 30 runs were carried out, each run evaluating 10,000 paths. I denotes the number of iterations (i.e. generations, in the case of the GA). The two rightmost columns show the average and estimated standard deviation of the resulting distribution of path lengths. See the main text for a complete description of the parameter settings for each method.

In order to allow a more fair comparison, a modified GA, denoted GA-NN, was therefore applied as well. In GA-NN, the chromosomes are generated by first selecting a random node and then, at each step, moving to the nearest neighbour. Next, in order to increase the diversity of the initial population to prevent premature convergence, the generated chromosomes are mutated by applying a few swap mutations. As is evident from the table, GA-NN easily outperforms GA-R. However, the ACO algorithms are still superior, particularly AS which generally finds paths with lengths less than 180 length units, quite near the length of the optimal path. MMAS could probably have done equally well, had one applied the alternating definition of the best candidate solution, described in Section 4.2.2. As a final point, though, one may also note that the GAs generally ran much faster than ACO, by around a factor 30. Thus, if the comparison had been based on equal running time, rather than equal number of evaluations, the difference in performance between ACO and GAs would certainly have been smaller.

Appendix A: Neural networks

A.1 Biological background

Artificial neural networks (ANNs) are inspired by the structure and properties of biological neural networks that constitute the brains of humans and animals. The brains of higher animals are, perhaps, the most complex objects that exist, and the computational capacity of such structures is truly amazing. Even though the silicon-based circuits of modern computers operate with a much higher clock frequency than the typical timescale of processes occurring inside the brains of animals, the number of operations per second of a computer is much smaller than that of a typical brain (of, say, a mammal). To a great extent, the superior computational capability of biological networks stems from the distributed nature of their computation: each such network consists of a very large number of interconnected basic elements, called neurons, that operate together to carry out the intricate computations needed for complex tasks such as image recognition, locomotion and other synchronized motions of the limbs, etc. Thus, we shall start with a brief description of neurons.

A.1.1 Neurons and synapses

A micrograph of a typical neuron is shown in the left panel of Fig. A.1. The right panel of the same figure shows a simplified, schematic depiction of a neuron. Starting from the cell body that contains the nucleus of the neural cell, a long filament known as the **axon** extends towards other neurons. This is the output wire of the neuron. At some distance from the cell body, the axon splits up into a delta of smaller wires ending on **synapses** that form the connections with other neurons. Input signals are received through the **dendrites** that derive their name from the Greek name for "tree" (*dendron*), a name that is obvious given their striking visual appearance, as seen in the left panel.

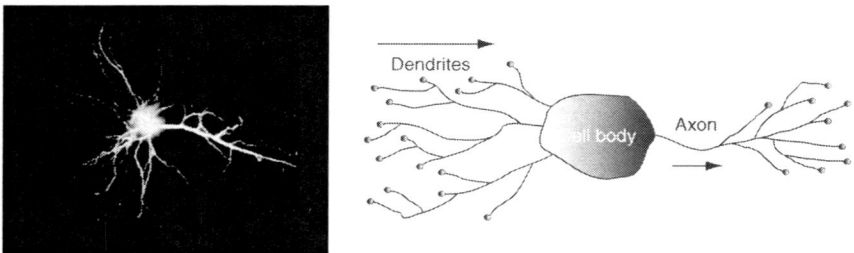

Figure A.1: *The left panel shows a micrograph of a neuron. Copyright: Richard Cole, Wadsworth Center, NYS. Dept. of Health, Albany, NY. Reproduced with permission. The right panel is a schematic representation of a neuron. The size of a neural cell body ranges from a few μm to around 100 μm.*

The signal flow within a neuron is electrical in nature: the signal consists of an electric potential (a **spike**) propagating through the axon with speeds of up to 100 m/s (see Fig. A.2). When the spike reaches a synapse connecting the axon to the dendrite of another neuron, the signal is transferred chemically. When the signal reaches the synapse, **transmitter substances** are released from the axonal (pre-synaptic) side, and diffuse towards the dendritic (post-synaptic) side, making the connection and triggering an electric potential that will propagate along the dendrite. Note that some synapses are, in fact, electrical rather than chemical. In such synapses, which occur in parts of the brain where very fast signal transfer is needed, the distance between the pre-synaptic and the post-synaptic sides is of order 3.5 nm [35], much shorter than the typical distance of around 30 nm for chemical synapses. The synapses connecting neurons can be either excitatory or inhibitory. An excitatory synapse increases the propensity of the post-synaptic neuron to fire, whereas an inhibitory synapse does the opposite. It should also be noted that neurons operate in a binary fashion: either a neuron fires a spike, or it does not.

After firing a spike, a neuron needs a period of recovery or relaxation, known as the **refractory period**, before a new spike can be fired. Typical refractory periods are of order 1 ms, thus limiting the firing frequency to around 1 kHz or less. As mentioned above, this can be compared with the clock frequencies of a few GHz – more than a million times faster – of typical contemporary computers.

A.1.2 Biological neural networks

Given the slow speed of neurons (compared to computers), how can the brain carry out such complex computational tasks as, for example, almost instantaneous facial recognition? The answer lies in the massively parallel architecture of the brain; the number of computational elements (cells) in a typical animal brain is indeed staggering: a rat has around 10^{10} neurons in its brain, whereas a human has around 10^{12} neurons and 10^{14}–10^{15} synapses. Thus, on average, each neuron is

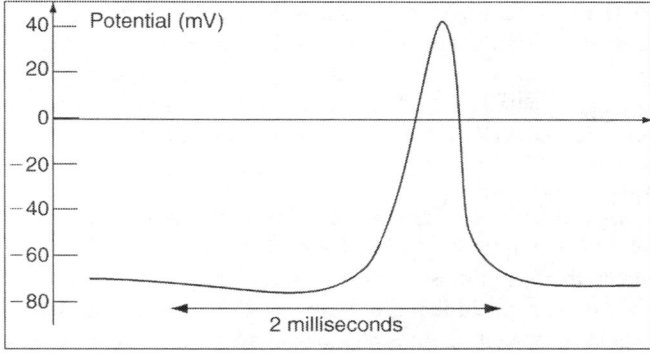

Figure A.2: *A typical neural spike.*

connected to hundreds or thousands of other neurons, forming an amazingly complex network with both short-range and long-range connections, and a multitude of feedback loops.

Thus, while a computer carries out its operations in a predominantly sequential manner (albeit very fast), the brain does its computations in parallel. Thus, for example, the approximately 1,000 billion neurons of the human brain are capable of carrying out around one million billion (10^{15}) operations per second. It should be noted here, however, that the analogy with computers should not be taken too far: the brain does not have the equivalent of a single, central clock that updates the neurons synchronously. Instead, it operates in a decentralized, asynchronous manner.

The computation carried out by a biological neural network (i.e. a brain) is determined by the synapses: if synapses are removed or added, the network architecture, and therefore its computation, changes. Thus, one may say that synapses are used for information storage, even though other properties, such as, for example, the timing between signals, may play an important role as well.

One of the most amazing features of animals, which will now be considered briefly, is their ability to memorize and learn, i.e. to store information about past events and to modify behaviour as a result of past experience.

A.1.3 Learning

As noted above, information is stored in synapses, and learning and memorization thus result from changes in synaptic strength. Since learning and memory are ubiquitous in animals, even in relatively simple ones as we shall see below, these two processes must provide an evolutionary advantage, and it is easy to understand, at least qualitatively, why this would be the case. Memory allows an animal to weigh past events when assessing the probable result of an action or a situation, and learning allows the animal to modify its behaviour, should the predicted outcome be negative. Long-term increases in synaptic strength, known as **long-term** potentiation (LTP) is a complex process that, on the molecular level, occurs in several phases, starting with **short-term potentiation** (STP) followed by early LTP and late LTP. Furthermore, synaptic weakening can occur as well, a process known as **long-term depression** (LTD). The relation between short-term memory (STM) and long-term memory (LTM) is further discussed below.

A.1.3.1 Hebbian learning

One of the first attempts to capture the mechanism behind learning was made by Hebb [29] in 1949. Essentially, Hebb stated that synapses will be strengthened between neurons that fire (almost) simultaneously. Thus, Hebb's hypothesis involves learning based on correlated firing of neurons. Following Hebb, letting x_i and x_j denote the signals propagating through the post-synaptic and pre-synaptic neurons, respectively, the strength w_{ij} of the synapse connecting these two neurons can be taken to change as

$$\frac{dw_{ij}}{dt} = \eta x_i x_j, \tag{A1}$$

where η is a parameter determining the learning rate. Of course, more complex formulations are possible as well, which take into account, for example, the timing of the signals through the two neurons.

Hebbian learning can be invoked to model, for example, the well-known classical conditioning experiment carried out by Pavlov [56], in which a dog was taught to associate the ringing of a bell with the presentation of food. Normally, the dog would salivate upon seeing food. If, however, the ringing of a bell precedes the presentation of food, the dog will eventually learn to associate the sound with the food, and will salivate even if no food is actually presented.

An obvious problem with Hebb's learning rule is that the synaptic strength cannot decrease. In other words, there is no mechanism for LTD. Hebb's rule can, however, be modified in various ways to include LTD as well. From biological experiments, it is known that low pre-synaptic activity coupled with high post-synaptic activity leads to synaptic weakening, an observation that suggests a modified Hebb rule of the kind

$$\frac{dw_{ij}}{dt} = \eta(x_i - \bar{x}_i)(x_j - \bar{x}_j), \tag{A2}$$

where $\bar{x}_i$ and $\bar{x}_j$ denotes the averages (over time) of x_i and x_j, respectively.

At the time when Hebb formulated his hypothesis, the state of neurophysiological research was not such that the hypothesis could be confirmed or refuted and, indeed, Hebb's postulate and the example regarding classical conditioning above are to a great extent phenomenological. However, empirical support for Hebb's hypothesis has been found, for example, in the study of memory storage in the brains of animals.

A.1.3.2 Habituation and sensitization

While Hebb's rule specifies a mechanism for learning, it does not state the exact molecular mechanism by which learning occurs. However, there have been studies, on a molecular level, of elementary forms of learning, such as **habituation** and **sensitization**. Habituation refers to the decrease in behavioural response (of an animal) to repeated, innocuous (neutral) stimuli, and it is a process found in a wide range of organisms, ranging from single-celled protozoa to humans. Habituation allows an organism to ignore irrelevant stimuli and thus to focus on stimuli important for survival. In view of its omnipresence, it apparently plays a very important role in the behavioural repertoire of animals. An example of habituation in humans is the gradual perceived fading of a ticking clock. Sensitization, by contrast, is an increase in behavioural response to a neutral stimulus following an aversive stimulus. An example, in humans, is the heightened state of alertness following the sound of a gunshot or an explosion.

Both habituation and sensitization have been studied extensively by Kandel [34] using the giant sea slug Aplysia, work that resulted in the Nobel Prize in Physiology or Medicine, in the year 2000. The choice of Aplysia as an object of study was motivated by the fact that this sea slug has very large and easily identifiable neurons. Furthermore, the number of neurons in Aplysia is only about 20,000. The left panel of Fig. A.3 shows an Aplysia specimen, and the right panel of the same figure shows a schematic drawing of the animal. The sea slug uses a gill for breathing, which is covered by a mantle and a shell. There is also a siphon, the function of which is to expel water from the animal. Aplysia is equipped with a basic defensive behaviour: if the siphon is stimulated, for example by a jet of water directed at it, the animal withdraws its gill under the protective cover of the mantle. However, if the stimulus is applied repeatedly, the response gradually becomes weaker. Aplysia also displays sensitization: an aversive stimulus, such as an electric shock applied to the tail of the animal, leads to an increased response to neutral stimuli.

Despite their apparent simplicity, both habituation and sensitization are, in fact, quite complex. A complete elucidation of the two processes must also distinguish between STM, lasting minutes to hours, and LTM, which can last for days or even the entire lifetime of the animal. Thus, several interesting questions present

Figure A.3: *Left panel: the sea slug Aplysia. Photo by Jason Dunsmore, reproduced with permission. Right panel: a schematic view of Aplysia. Copyright The Nobel Assembly at Karolinska Institutet, reproduced with permission.*

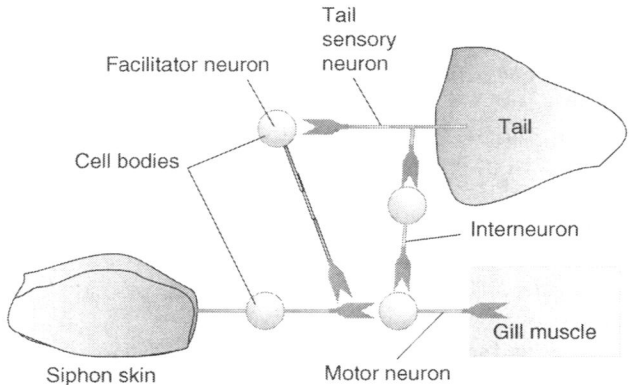

Figure A.4: *A schematic illustration of the neural circuit in Aplysia responsible for habituation and sensitization. The circuit consists of two sensory neurons connected to the siphon and the tail, respectively, a motor neuron innervating muscles under the gill, and a facilitator interneuron.*

themselves in connection with the study of habituation and sensitization in Aplysia: how are memories formed? What processes distinguish short-term and long-term retention of memories? In order to answer these questions, Kandel carried out a series of detailed experiments, through which the neural circuit responsible for habituation and sensitization was uncovered. A schematic illustration of the circuit is shown in Fig. A.4. In addition, the experiments showed that, in both STM and LTM, the variation in behavioural response (in habituation and sensitization) results from a change in neurotransmitter release whereas LTM also depends on anatomical changes. LTM is elicited in response to repeated stimuli, which trigger a cascade of reactions that, ultimately, lead to a change in gene expression resulting in the growth of new synapses. Thus, LTM depends on what Kandel referred to as *a dialogue between genes and synapses* [34]. Having briefly introduced the components and structure of biological neural networks, as well as some elementary forms of learning, we will now turn to the artificial counterpart.

A.2 Artificial neural networks

There exists many different types of ANNs, and the taxonomy of such networks can be constructed in different ways. A common approach is to classify ANNs based on their structure or, more exactly, based on the presence or absence of feedback connections in the networks. Thus, a **feedforward neural network** (FFNN) contains only forward-pointing connections, whereas a **recurrent neural network** (RNN) contains both forward-pointing and backward-pointing (feedback) connections. Each connection is associated with a **connection weight** (or, simply, **weight**) with a certain numerical value. The connection weights correspond to the synapses

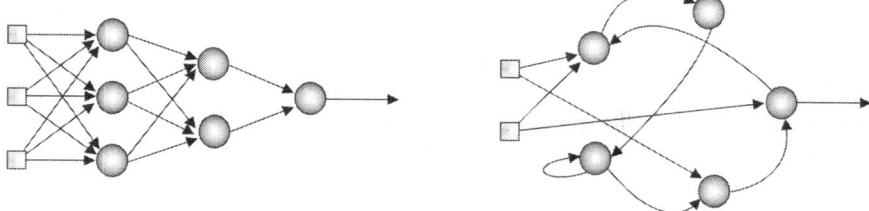

Figure A.5: *A layered feedforward neural network (FFNN, left panel) and a recurrent neural network (RNN, right panel). The computational elements (neurons), further described in Section A.2.1, are shown as circles. The squares are input elements that simply mediate signals without carrying out any further computation.*

in biological neurons. The two basic types of ANNs are illustrated in Fig. A.5. As can be seen in the left panel of the figure, FFNNs are normally layered, with a layer of **input elements** (always shown as squares, in this book) that simply mediate the signals, without carrying out any further computation, to the neurons (always shown as discs) in the first **hidden layer** (there may be more than one) that, after carrying out an elementary operation that will be described below, in turn propagate the signals to the next hidden layer, etc., until, finally, the **output layer** is reached and the output of the network thus can be computed. ANNs can also be classified on the basis of the temporal aspects of their computation. Here, the two main classes are discrete-time neural networks (DTNNs, the most common kind) and continuous-time neural networks (CTNNs). Networks operating in discrete time are useful, for example, in applications involving function approximation or data classification, whereas CTNNs are employed, for instance, in control applications where a continuously varying signal is needed.

There are also numerous methods for training neural networks, i.e. for setting the weights in the network (and, more rarely, also its structure) so as to make it solve a desired computational task. Typically, learning methods are divided into supervised and unsupervised methods, even though some methods do not fall cleanly into either category. In **supervised methods**, for each input vector presented to the ANN, there must be a desired output vector. Thus, in such learning methods, the objective is to minimize the difference (the error) between the desired network output and the actual output, the former acting as a teacher guiding the learning process. By contrast, **unsupervised learning methods** are applied in cases where one cannot generate an error signal for each input vector presented to the network. The term *unsupervised* is, however, something of a misnomer: even in this case, some form of feedback is normally given to the ANN, even though it is not given at every time step but, perhaps, only at the end of some lengthy evaluation. Thus, we shall introduce two novel terms: supervised learning will henceforth be referred to as **strongly guided learning** and unsupervised learning will be referred to as **weakly guided learning**.

As mentioned in the Preface, ANNs are not an optimization method *per se*. Instead, they are computational structures to which several learning (optimization) methods can be applied, some of which we shall now briefly consider. One should, however, keep in mind that, in addition to the dedicated neural network methods described below, the various stochastic optimization algorithms (e.g. PSO) considered in this book can *also* be applied for the training of neural networks.

A.2.1 Artificial neurons

As in the biological counterpart, the basic computational elements of ANNs are referred to as neurons. Several different artificial neuron models have been developed. Here, we shall only consider one of the most common, namely the **McCulloch–Pitts** (**MCP**) **neuron** [48], illustrated in Fig. A.6. In the MCP neuron, the output signal y is generated by first computing a weighted sum of the inputs

$$s = \sum_{j=1}^{n} w_j x_j + b, \tag{A3}$$

where w_j are connection weights, x_j are the input signals (e.g. output signals from other neurons), and b is the **bias term**. In order to simplify the notation, the bias term is often written $w_0 x_0$, where $w_0 = b$ and x_0 is always 1. Thus, s can then be expressed as

$$s = \sum_{j=0}^{n} w_j x_j. \tag{A4}$$

Next, the neuron output is obtained as

$$y = \sigma \left(\sum_{j=0}^{n} w_j x_j \right), \tag{A5}$$

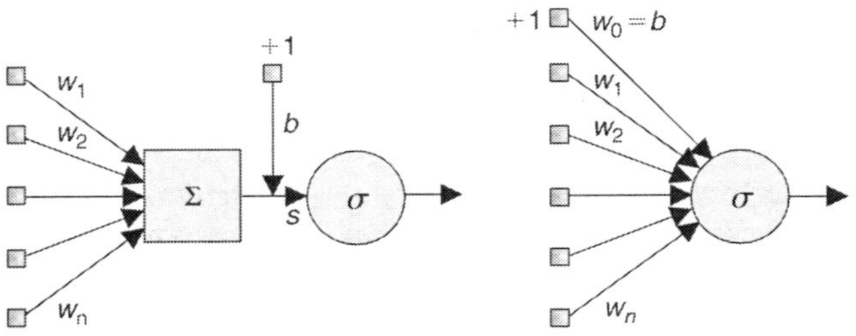

Figure A.6: *The McCulloch–Pitts neuron (left panel), and the simplified notation normally used (right panel).*

where σ is the **activation function**.[1] Now, as mentioned above, biological neurons have the property of all-or-nothing firing, something that can be modelled using an activation function that takes the value 1 if s exceeds some threshold T, and 0 otherwise. In that case, the bias term determines the propensity of the neuron to fire in the absence of any input. However, while this model is biologically plausible (albeit strongly simplified), it is not always suitable in applications, where a graded response (i.e. a number in the *range* [0, 1]) often is more appropriate. Thus, it is common instead to use the **logistic sigmoid**

$$\sigma(s) = \frac{1}{1+e^{-cs}}, \tag{A6}$$

(where c is a positive constant), as the activation function. One may still retain a biological interpretation, by considering the output y from the neuron as the *average* firing frequency over some time interval, rather than the actual output spike of the neuron.

The neuron model considered above is intended for applications in discrete time. It can, however, easily be generalized for use in continuous-time applications, by the addition of the time constant τ. The model then takes the form

$$\tau \frac{dy}{dt} + y = \sigma \left(\sum_{j=0}^{n} w_j x_j \right). \tag{A7}$$

A.2.2 Feedforward neural networks and backpropagation

The most common type of ANN (in applications) is the FFNN, illustrated in the left panel of Fig. A.5, and it is most commonly used in connection with strongly guided learning methods. As mentioned above, FFNNs are layered, such that neurons in one layer connect only to the neurons in the next layer.

A.2.2.1 The Delta rule

To begin with, consider the simplest possible form of ANN, consisting of a single neuron (and, therefore, a single layer of network weights). For simplicity, let us use a linear activation function $\sigma(s) = s$, so that

$$y = \sum_{j=0}^{n} w_j x_j. \tag{A8}$$

---

[1] The activation function is also sometimes referred to as a **squashing function** or a **sigmoid function**.

Now, assume also that a set of training data, consisting of M **input–output pairs**,[2] is available. Thus, for each member m of this set, an input vector

$$\mathbf{x}^{[m]} = \left(x_1^{[m]}, x_2^{[m]}, \ldots, x_n^{[m]}\right)^{\mathrm{T}}, \tag{A9}$$

is defined, along with a desired (scalar) output $o^{[m]}$. Thus, if input vector m is applied to this single-neuron ANN, an error $e^{[m]}$ can be formed as

$$e^{[m]} = o^{[m]} - y^{[m]} = o^{[m]} - \sum_{j=0}^{n} w_j x_j^{[m]}. \tag{A10}$$

Given the errors $e^{[m]}$ for each element of the training set, the total squared error E can be formed as

$$E = \frac{1}{2} \sum_{m=1}^{M} (e^{[m]})^2 = \frac{1}{2} \sum_{m=1}^{M} \left(o^{[m]} - \sum_{j=0}^{n} w_j x_j^{[m]} \right)^2. \tag{A11}$$

For any given training set, the variables in eqn (A11) are the connection weights w_j, and it is evident from the equation that the error E is a quadratic function of the weights. The objective, of course, is to minimize the error, and the simplest way of doing so is to apply gradient descent (see Section 2.4.1), i.e. to modify the weights according to

$$w_j \leftarrow w_j + \Delta w_j, \tag{A12}$$

where

$$\Delta w_j = -\eta \frac{\partial E}{\partial w_j}, \tag{A13}$$

where $\partial E / \partial w_j$ are the components of the gradient

$$\nabla_w E = \left(\frac{\partial E}{\partial w_0}, \frac{\partial E}{\partial w_1}, \ldots, \frac{\partial E}{\partial w_n} \right)^{\mathrm{T}}. \tag{A14}$$

---

[2] The input–output pairs are also referred to as **training elements**.

Using eqn (A11), the partial derivative of the error E can be computed as

$$\frac{\partial E}{\partial w_j} = \frac{1}{2}\frac{\partial}{\partial w_j}\sum_{m=1}^{M}(e^{[m]})^2 = \sum_{m=1}^{M} e^{[m]}\frac{\partial e^{[m]}}{\partial w_j}$$

$$= \sum_{m=1}^{M} e^{[m]}\frac{\partial\left(o^{[m]} - \sum_{k=0}^{n} w_k x_k^{[m]}\right)}{\partial w_j} = -\sum_{m=1}^{M} e^{[m]} x_j^{[m]}$$

$$= -\sum_{m=1}^{M}(o^{[m]} - y^{[m]})x_j^{[m]}. \qquad (A15)$$

Thus, the weight change Δw_j takes the form

$$\Delta w_j = -\eta\frac{\partial E}{\partial w_j} = \eta\sum_{m=1}^{M}(o^{[m]} - y^{[m]})x_j^{[m]}. \qquad (A16)$$

This rule is known as the **Delta rule** or the **Widrow–Hoff rule** [79] in honour of its discoverers.

A.2.2.2 Limitations of single-layer networks
The Delta rule is easy to apply but, unfortunately, there are limits on the representational power of single-layer neural networks. In fact, it is simple to show that in, for example, a binary classification task (i.e. with two classes), a single-layer network can only represent data sets that are linearly separable. However, as we shall see below, multi-layer ANNs can represent more complex functions, provided that the activation function is non-linear.[3] We shall now derive a strongly guided learning method, known as backpropagation, for such networks.

A.2.2.3 Backpropagation
Consider the network shown in Fig. A.7. It is assumed that there is a training data set for which the desired output vector $\mathbf{o}^{[m]}$, $m = 1,\ldots,M$ is known for each input vector $\mathbf{x}^{[m]}$. For any input vector $\mathbf{x} = \mathbf{x}^{[m]}$, the corresponding output vector $\mathbf{y}^O$ can be computed and an error signal $\mathbf{e}^{[m]}$ whose components $e_l = e_l^{[m]}$ are defined as

$$e_l = o_l - y_l^O, \qquad (A17)$$

can be computed for each neuron l, $l = 1,\ldots,n_O$, where n_O is the number of neurons in the output layer. In eqn (A17), the term o_l denotes the lth component of the desired output vector $\mathbf{o} = \mathbf{o}^{[m]}$, and in most of the equations that will follow in this section, the index m that enumerates the input–output pairs will be dropped. In fact, even

---

[3] A multi-layer neural network with a *linear* activation function is equivalent to a single-layer network.

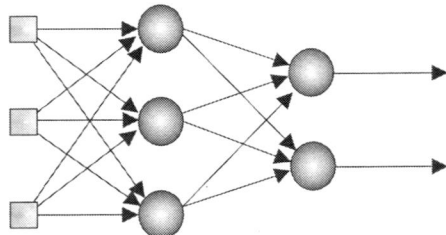

Figure A.7: *A multilayer feedforward neural network. Note that the biases are not shown in this figure.*

though, as in eqn (A16), the training of the network can be based on the errors obtained over the entire data set, a procedure known as **batch training**, it is usually very time-consuming to do so: each weight modification will then require that one should run through the entire data set. Instead, it is common to use **stochastic gradient descent**, in which the elements of the training data set are considered one by one, in random order. Thus, in stochastic gradient descent, an approximate error hypersurface is used, based on a single input–output pair. A random permutation of the M training data set elements is formed and, for each element, a weight update is carried out. For a given element m, the total error can be defined as

$$\mathcal{E}^{[m]} \equiv \mathcal{E} = \frac{1}{2} \sum_{l=1}^{n_O} e_l^2. \tag{A18}$$

Thus, for an output neuron, the corresponding error signal can easily be computed. But what about the neurons in the *hidden* layer(s)? For these neurons, whose output are denoted $\mathbf{y}^H$, no simple error signal can be formed, since it is not obvious how to reward (or punish) a hidden neuron for a result that appears in the output layer. We are therefore faced with a **credit assignment problem** for the neurons in the hidden layer(s). However, let us begin by considering the weights connecting the hidden layer to the output layer. For a neuron i in the output layer, the partial derivative of the error $\mathcal{E}$ with respect to the weight $w_{ij}^{H \to O}$, connecting neuron j in the hidden layer to neuron i (see Fig. A.8), can be written

$$\frac{\partial \mathcal{E}}{\partial w_{ij}^{H \to O}} = \frac{\partial}{\partial w_{ij}^{H \to O}} \left(\frac{1}{2} \sum_{l=1}^{n_O} e_l^2 \right) = e_i \frac{\partial e_i}{\partial w_{ij}^{H \to O}}$$

$$= e_i \frac{\partial}{\partial w_{ij}^{H \to O}} \left(o_i - \sigma \left(\sum_{s=0}^{n_H} w_{is}^{H \to O} y_s^H \right) \right)$$

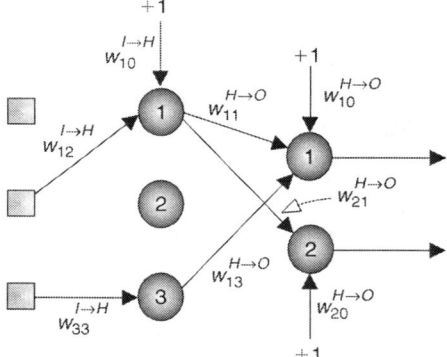

Figure A.8: *The notation used in the derivation of the backpropagation rule, here shown for the case $n_I = 3, n_H = 3, n_O = 2$. $w_{ij}^{H \to O}$ is the weight connecting neuron j in the hidden layer to neuron i in the output layer. Similarly, $w_{ij}^{I \to H}$ connects input element j to neuron i in the hidden layer. For clarity, only a few of the available connections have been drawn in the figure.*

$$= -e_i \sigma' \left(\sum_{s=0}^{n_H} w_{is}^{H \to O} y_s^H \right) \frac{\partial \left(\sum_{s=0}^{n_H} w_{is}^{H \to O} y_s^H \right)}{\partial w_{ij}^{H \to O}}$$
$$= -e_i \sigma' y_j^H, \tag{A19}$$

where n_H denotes the number of neurons in the hidden layer and, in the last step, the argument of the derivative of the activation function σ has been omitted, for brevity. In analogy with the Delta rule, eqn (A16), the weight modifications are now given by

$$\Delta w_{ij}^{H \to O} = -\eta \frac{\partial \mathcal{E}}{\partial w_{ij}^{H \to O}} = \eta \delta_i y_j^H, \tag{A20}$$

where $\delta_i = e_i \sigma'$ is the **local gradient**.

Consider now a neuron in the hidden layer in Fig. A.7 or Fig. A.8, with output signal y_i^H. Since this neuron is connected to all the neurons in the output layer, via the connection weights $w_{ki}^{H \to O}, k = 1, \ldots, n_O$, it will give a contribution to the error of all those neurons. Formally, we can also in this case write the weight modification rule as

$$\Delta w_{ij}^{I \to H} = -\eta \frac{\partial \mathcal{E}}{\partial w_{ij}^{I \to H}}. \tag{A21}$$

For this expression to be useful, we must compute the partial derivative of $\mathcal{E}$ with respect to $w_{ij}^{I \to H}$. Proceeding in a way similar to that used for the output neurons considered above, and noting that only y_i^H depends on $w_{ij}^{I \to H}$, the chain rule of differentiation gives

$$\frac{\partial \mathcal{E}}{\partial w_{ij}^{I \to H}} = \frac{\partial \mathcal{E}}{\partial y_i^H} \frac{\partial y_i^H}{\partial w_{ij}^{I \to H}} = \frac{\partial \mathcal{E}}{\partial y_i^H} \frac{\partial}{\partial w_{ij}^{I \to H}} \sigma \left(\sum_{p=0}^{n_I} w_{ip}^{I \to H} y_p^I \right)$$

$$= \frac{\partial \mathcal{E}}{\partial y_i^H} \sigma' \left(\sum_{p=0}^{n_I} w_{ip}^{I \to H} y_p^I \right) y_j^I, \qquad (A22)$$

where, for the two-layer network, y_p^I denotes the signal coming from input element p, i.e. $y_p^I = x_p$, and n_I denotes the number of input elements. Continuing the calculation, we get

$$\frac{\partial \mathcal{E}}{\partial y_i^H} = \frac{\partial}{\partial y_i^H} \left(\frac{1}{2} \sum_{l=1}^{n_O} e_l^2 \right) = \sum_{l=1}^{n_O} e_l \frac{\partial e_l}{\partial y_i^H} = \sum_{l=1}^{n_O} e_l \frac{\partial \left(o_l - y_l^O \right)}{\partial y_i^H}$$

$$= -\sum_{l=1}^{n_O} e_l \frac{\partial}{\partial y_i^H} \sigma \left(\sum_{s=0}^{n_H} w_{ls}^{H \to O} y_s^H \right)$$

$$= -\sum_{l=1}^{n_O} e_l \sigma' \left(\sum_{s=0}^{n_H} w_{ls}^{H \to O} y_s^H \right) w_{li}^{H \to O}$$

$$= -\sum_{l=1}^{n_O} \delta_l w_{li}^{H \to O}, \qquad (A23)$$

where, in the final step, δ_l is defined as in eqn (A20). Combining eqns (A21), (A22), and (A23), the weight modification can be written

$$\Delta w_{ij}^{I \to H} = \eta \kappa_i y_j^I, \qquad (A24)$$

where

$$\kappa_i = \sigma' \left(\sum_{p=0}^{n_I} w_{ip}^{I \to H} y_p^I \right) \sum_{l=1}^{n_O} \delta_l w_{li}^{H \to O}. \qquad (A25)$$

From eqn (A24) it is clear that the expression for the weight modification in the hidden layer is similar to the corresponding expression for the output layer. The main difference is that δ is replaced by κ which, in turn, contains a weighted sum of the δ-terms obtained for the output layer.

The weight modification rules defined in eqns (A20) and (A24) are the central components of the backpropagation algorithm. It derives its name from the fact that errors are propagated backward in the network; in order to compute the weight

1. Initialize the network by assigning small, random values to all weights $w_{ij}^{H \to O}, j = 0, \ldots, n_H, i = 1, \ldots, n_O$ and $w_{ij}^{I \to H}, j = 0, \ldots, n_I, i = 1, \ldots, n_H$.
2. Generate a random permutation of the M input–output pairs.
3. For each input–output pair in this permutation, do the following:

    3.1. Compute the output $\mathbf{y}^O$ of the network, and form the error $\mathbf{e}$ between the desired output and the actual output.
    3.2. Compute $\Delta w_{ij}^{H \to O}$ according to eqn (A20).
    3.3. Compute $\Delta w_{ij}^{I \to H}$ according to eqn (A24).
    3.4. Update the weights according to

    $$w_{ij}^{H \to O} \leftarrow w_{ij}^{H \to O} + \Delta w_{ij}^{H \to O}$$

    and

    $$w_{ij}^{I \to H} \leftarrow w_{ij}^{I \to H} + \Delta w_{ij}^{I \to H}$$

4. Every K^{th} epoch, compute the error over the whole training set, according to eqn (A26).
5. Repeat steps 2–4 until the error drops to the desired level.

Algorithm A.1: *The backpropagation algorithm, for the case of an FFNN with a single hidden layer. Note that the range of the indices* ij *have been written explicitly only in step 1. However, the same ranges are implied for all subsequent steps.*

change in the hidden layer, the local gradient δ from the output layer must first be known.

Note that the modifications Δw_{ij} are computed for *all* weights in the network before *any* modifications are actually applied. Thus, in the computation of the modifications of the weights $w_{ij}^{I \to H}$, the old values are used for the weights $w_{ij}^{H \to O}$.

In the derivation above, we limited ourselves to the case of two-layer networks. It is easy to realize, though, that the formulae for the weight changes can be used even in cases where there are more than one hidden layer. The weight modification in the first hidden layer (i.e. the one that follows immediately after the input elements) would then be computed as $\eta \gamma_i y_j^I$, where the γ_i are computed using the κ_i which, in turn, are obtained from the δ_i.

When the backpropagation algorithm is used, weight modifications are typically carried out for each of the M input–output pairs in the training set. The M weight updates carried out during a loop through the entire training data set is referred to as a training **epoch**. As mentioned above, the elements of the training set are considered in random order. Thus, at the beginning of each training epoch, a random permutation of the numbers $1, \ldots, M$ is formed, determining the order in which the elements of the training set are to be considered. The progress of the algorithm is commonly measured by computing, at the end of every K^{th} training epoch, the RMS error

$$E_{\text{RMS}} = \sqrt{\frac{1}{M}\sum_{m=1}^{M} 2\mathcal{E}^{[m]}}, \quad (A26)$$

where the factor 2 has been introduced to remove the factor 1/2 included, for convenience, in eqn (A18). With this definition, E_{RMS} measures the difference, for a typical input–output pair, between the desired output and the actual output. Note that a true measure of the error requires that the computation should be carried out with fixed weights (hence, at the *end* of a training epoch); even though it is possible to form a complete error over the training set by summing the errors obtained for each of the M training elements during a training epoch, the error thus formed would not represent the true error, since the weights are constantly being modified (once for each input–output pair).

At the start of a backpropagation run, the weights are first initialized to small, random values. The speed by which learning progresses is dependent on the value of the learning rate parameter η. If η is chosen too large, the error will jump around wildly on the error surface. On the other hand, if η is chosen too small, convergence towards the minimum error will be stable but very slow. The problem of choosing a value of η is made even more difficult by the fact that the optimal value of this parameter will not stay constant during a run. Initially, when the weights are randomly assigned, there are many directions in weight space leading to a reduction in the error so that a large value of η can be used. As the training progresses, and the error E_{RMS} approaches its minimum, smaller values of η should be used. In the literature, several different methods for selecting a functional form for the variation of η have been suggested. A common choice is to let η vary with the training epoch (denoted k) according to

$$\eta(k) = \eta_0 \frac{1 + \frac{c}{\eta_0}\frac{k}{T}}{1 + \frac{c}{\eta_0}\frac{k}{T} + T\frac{k^2}{T^2}}. \quad (A27)$$

Using this equation, a near-constant $\eta \approx \eta_0$ is obtained for small k. For $k \gg T$, η varies as c/k. The backpropagation algorithm is summarized in Algorithm A.1. We shall now consider two examples, one analytical example involving a computation of the weight modifications in a single backpropagation step, and one numerical example.

Example A.1
Carry out a single backpropagation step for the network shown in Fig. A.9, given that the input $(x_1, x_2)^T$ equals $(1, 0)^T$, and that the desired output o_1 equals 1. The learning rate η equals 1, and the activation function is the logistic sigmoid with $c = 1$.

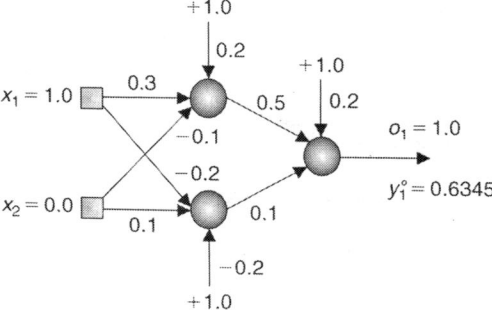

Figure A.9: *The neural network considered in Example A.1.*

Solution Let us first compute the error. The output from the neurons in the hidden layer becomes

$$y_1^H = \sigma\left(\sum_{p=0}^{2} w_{1p}^{I \to H} y_p^I\right) = \sigma(0.2 \times 1.0 + 0.3 \times 1.0 - 0.1 \times 0)$$

$$= \sigma(0.5) = \frac{1}{1 + e^{-0.5}} = 0.6225, \tag{A28}$$

$$y_2^H = \sigma\left(\sum_{p=0}^{2} w_{2p}^{I \to H} y_p^I\right) = \sigma(-0.2 \times 1.0 - 0.2 \times 1.0 + 0.1 \times 0)$$

$$= \sigma(-0.4) = \frac{1}{1 + e^{0.4}} = 0.4013. \tag{A29}$$

Similarly, the output of the neuron in the output layer becomes

$$y_1^O = \sigma\left(\sum_{s=0}^{2} w_{1s}^{H \to O} y_s^H\right) = \sigma(0.2 \times 1.0 + 0.5 \times 0.6225 + 0.1 \times 0.4013)$$

$$= \sigma(0.5514) = 0.6345. \tag{A30}$$

Using eqn (A17), the error signal can be computed as

$$e_1 = o_1 - y_1^O = 1 - 0.6345 = 0.3655. \tag{A31}$$

The derivative of the activation function is given by

$$\sigma'\left(\sum_{s=0}^{2} w_{1s}^{H \to O} y_s^H\right) = \sigma'(0.5514) = 0.2319. \tag{A32}$$

Thus,

$$\delta_1 = e_1 \sigma'\left(\sum_{s=0}^{2} w_{1s}^{H \to O} y_s^H\right) = 0.3655 \times 0.2319 = 0.0848. \tag{A33}$$

Using eqn (A20), the modification of the weight $w_{10}^{H \to O}$ is obtained as

$$\Delta w_{10}^{H \to O} = \eta \delta_1 y_0^H = 1 \times 0.0848 \times 1 = 0.0848. \tag{A34}$$

Using the same procedure, the change in the weights $w_{11}^{H \to O}$ and $w_{12}^{H \to O}$ are obtained

$$\Delta w_{11}^{H \to O} = \eta \delta_1 y_1^H = 1 \times 0.0848 \times 0.6225 = 0.0528, \tag{A35}$$

$$\Delta w_{12}^{H \to O} = \eta \delta_1 y_2^H = 1 \times 0.0848 \times 0.4013 = 0.0340. \tag{A36}$$

Proceeding now to the hidden layer, we first compute κ_1 as follows

$$\kappa_1 = \sigma' \left(\sum_{p=0}^{2} w_{1p}^{I \to H} y_p^I \right) \sum_{l=1}^{1} \delta_l w_{l1}^{H \to O}$$

$$= \sigma'(0.5) \delta_1 w_{11}^{H \to O} = 0.2350 \times 0.0848 \times 0.5 = 0.00996. \tag{A37}$$

The change in the weight $w_{10}^{I \to H}$ is now given by

$$\Delta w_{10}^{I \to H} = \eta \kappa_1 y_0^I = 1 \times 0.00996 \times 1 = 0.00996. \tag{A38}$$

The modifications to the weights $w_{11}^{I \to H}$ and $w_{12}^{I \to H}$ are computed in the same way, resulting in (check!)

$$\Delta w_{11}^{I \to H} = 0.00996, \tag{A39}$$

$$\Delta w_{12}^{I \to H} = 0. \tag{A40}$$

Using the same method, κ_2 can be obtained, and thereby also the modifications of the weights entering the second neuron in the hidden layer. The result is (check!)

$$\Delta w_{20}^{I \to H} = 0.00204, \tag{A41}$$

$$\Delta w_{21}^{I \to H} = 0.00204, \tag{A42}$$

$$\Delta w_{22}^{I \to H} = 0. \tag{A43}$$

With the updated weights, the output of the network becomes, using the same input signal $(x_1 = 1, x_2 = 0)$

$$y_1^O = 0.6649. \tag{A44}$$

Thus, with the new weights, the error is reduced from 0.3655 to

$$e_1 = o_1 - y_1^O = 1 - 0.6649 = 0.3351. \tag{A45}$$

■

Table A.1: *The truth table for the 3-parity function, see Example A.2.*

x_1	x_2	x_3	o	x_1	x_2	x_3	o
0	0	0	0	1	0	0	1
0	0	1	1	1	0	1	0
0	1	0	1	1	1	0	0
0	1	1	0	1	1	1	1

Example A.2

An *n*-**parity function** takes n (binary-valued) inputs and outputs a 1 if an odd number of inputs are equal to 1, and zero otherwise. Use backpropagation to train an FFNN so that it can represent the 3-parity function.

Solution For $n = 3$ there are eight input–output pairs as shown in Table A.1. Since there are three inputs, a 3-3-1 network[4] was chosen. Backpropagation with stochastic gradient descent was applied, in each epoch generating a random permutation of the eight input–output pairs and carrying out an equal number of backpropagation steps, one for each input–output pair. At the end of every epoch the RMS training error

$$E_{\text{RMS}} = \sqrt{\frac{1}{8}\sum_{m=1}^{8}(\epsilon^{[m]})^2}, \tag{A46}$$

was measured, where $\epsilon^{[m]}$ is the error for input–output pair m, i.e. the difference $o - y^O$ between the desired output and the actual output. The weights were initialized randomly in the range $[-0.2, 0.2]$. The training parameter η was set to 0.3, and was kept constant throughout the run. The activation function for the neurons was taken as the standard sigmoid defined in eqn (A6), with $c = 2$. The results are shown in Fig. A.10. As can be seen from the figure, after an initial transient, the error drops from around 0.5 to a plateau around 0.3, before finally dropping towards zero. ∎

A.2.3 Recurrent neural networks

While FFNNs can be made to represent a great variety of input–output mappings, they suffer from one fundamental drawback: regardless of the values of the network weights, the output from an FFNN will always be the same for any given input signal. In other words, FFNNs lack dynamic memory. Clearly, dynamic memory is an important property in biological systems, allowing an animal to respond differently on a second occurrence of a given situation, should the first response be inappropriate. In order to provide ANNs with dynamic memory, recurrent (feedback) connections may be introduced. Several different types of such recurrent

---

[4] In general, an FFNN with n_I input elements, n_H hidden neurons, and n_O output neurons is denoted an n_I-n_H-n_O network.

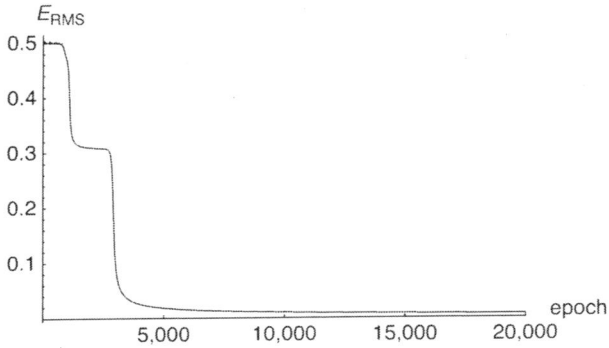

Figure A.10: *The RMS error over the training set for the 3–parity problem, plotted as a function of the training epoch, see Example A.2.*

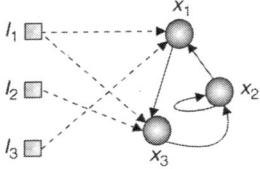

Figure A.11: *A continuous-time recurrent neural network (CTRNN). The input weights w_{ij}^I and weights w_{ij} are shown as dashed and solid lines, respectively. Note that not all possible weights are present in this particular network; only the input weights $w_{11}^I, w_{13}^I, w_{31}^I$ and w_{32}^I and the weights w_{12}, w_{22}, w_{23} and w_{31} have non-zero values. The biases are not shown.*

neural networks (RNNs) have been suggested in the literature, for example, **Jordan networks**, in which the output of the network is fed back to the input layer, and **Elman networks**, in which there are feedback connections from the hidden units to the input layer. The most general type of RNNs, however, are those that are fully recurrent, i.e. where any neuron may, in principle, receive input from any other neuron (including itself) and from any input element.

Furthermore, in certain cases, for example, many control applications, a smoothly varying output signal may be needed, i.e. one that does not exhibit discontinuous jumps even if the input to the network should do so. In such applications, it is suitable to use a **continuous-time recurrent neural network** (CTRNN), an example of which is shown in Fig. A.11. In CTRNNs the output of neuron i varies according to

$$\tau_i \frac{dx_i}{dt} + x_i = \sigma \left(\sum_{j=1}^{n} w_{ij} x_j + \sum_{j=1}^{m} w_{ij}^I I_j + b_i \right), \qquad (A47)$$

where n and m are the number of neurons and the number of input elements, respectively, τ_i are time constants and b_i the bias terms. I_j are the input signals. The weights w_{ij} connect neuron j to neuron i, and the (input) weights w^I_{ij} connect input element j to neuron i. As in the case of discrete-time ANNs, there are many alternatives for the activation function σ, the logistic sigmoid being the most common in cases where the signals x_i are to be constrained to the interval $[0, 1]$. In cases where negative outputs are needed, a common choice of activation function is

$$\sigma(s) = \tanh(cs). \tag{A48}$$

Even though the network described by eqn (A47) operates in continuous time, in practical applications the frequency of input signal updates is, of course, limited. Thus, assuming a time step Δt, the discretized version of eqn (A47) becomes, after some elementary re-arrangement of terms,

$$x_i(t + \Delta t) = x_i(t) + \frac{\Delta t}{\tau_i} \left[-x_i(t) + \sigma \left(\sum_{j=1}^{n} w_{ij} x_j(t) + \sum_{j=1}^{m} w^I_{ij} I_j(t) + b_i \right) \right] \tag{A49}$$

where the time dependence of x_i and I_j has been made explicit. Given $I_j(t)$ and initial values $x_i(0)$, the network can thus easily be integrated. Note that, in order for the discretization to be valid, all time constants τ_i must fulfil

$$\tau_i \gg \Delta t. \tag{A50}$$

In cases where a desired output can be specified for each input signal, a CTRNN can be trained using **recurrent backpropagation** which, as the name implies, is a generalization of the backpropagation algorithm [28]. However, it is not always the case that a desired output can be specified. For example, consider a problem involving the motion of a differentially steered two-wheeled autonomous robot, where the output from two of the neurons x_L and x_R in the CTRNN, properly scaled, may be used for specifying the torques applied to the left and right wheel of the robot, respectively; here, the result of applying a given signal may not become apparent until a later time. Furthermore, in applications of the kind just mentioned, it may also be difficult to specify the appropriate number of neurons (n) in advance.

In applications where either (or both) of the conditions just mentioned apply, the training must be carried out using other methods, for example the stochastic optimization methods considered in Chapters 3 through 5.

A.2.4 Other networks

In addition to the layered FFNNs and the CTRNNs considered above, many other kinds of ANNs have been introduced as well. Examples include **radial-basis function networks** [28], which are typically used in, for example, time series prediction and function approximation, **Kohonen maps**, [38] also known as **self-organizing** feature maps (SOFMs) that are used for forming low-dimensional (normally two-dimensional) representations of multidimensional data, and **Hopfield networks** [33], which are used for memory storage, and have the important features of error correction and memory completion, i.e. the ability to generate the correct output even when the input is garbled or incomplete.

A.3 Applications

ANNs have been used in a wide range of applications, and new areas of application are constantly appearing. A few examples of the use of ANNs are given in Appendix C and in Chapter 6 where different optimization methods are compared. Here, only a brief list of other applications will be given. Some of the most common applications of ANNs include function approximation, pattern recognition, prediction and forecasting, classification, and control. In function approximation, which is typically strongly guided using the terminology introduced above, an ANN is trained to represent a function $f = f(\mathbf{x})$ for any $\mathbf{x}$ in the domain of f. Normally, the training samples cover only a subset of the domain, and the ANN is relied upon to provide the correct output for intermediate points.

Pattern recognition comes in many different forms, for example, speech recognition, hand-written digit recognition, identification of faces, etc., and typically involves strongly guided training of an FFNN. Prediction and forecasting plays an important role in, for example, finance, macroeconomics, and meteorology, and it may involve both FFNNs and RNNs. In classification tasks, the aim is to divide samples into two or more classes, based on a set of features measured for each sample. While many other methods exist for classification (such as, for example, k-nearest neighbour classifiers (see Section 3.6.3), linear discriminant analysis, etc.), ANNs have also been applied, for example in medical classification tasks, where the aim often is to discover which combination of features (if any) is indicative of a given disease. Thus, in such applications, the set of samples consists of measurements both from individuals who have the disease in question and from a control group. Control applications of ANNs include, for example, the control of production facilities, process control and the control of vehicles and robots (of both the stationary and autonomous kinds).

Appendix B: Analysis of optimization algorithms

B.1 Classical optimization

B.1.1 Global minima of convex functions

Theorem In a convex optimization problem, i.e. a problem of the form

$$\text{minimize } f(\mathbf{x}), \quad \mathbf{x} \in \mathbf{S}, \tag{B1}$$

where $f(\mathbf{x})$ is a convex function, and $\mathbf{S} \in \mathbf{R}^n$ is a convex set, every local minimum is also a global minimum.

Proof Assume that $\mathbf{x}^*$ is a local minimum which is *not* the global minimum, i.e. a minimum such that, contrary to the assertion above, there exists an $\mathbf{x}^\dagger \neq \mathbf{x}^* \in \mathbf{S}$ such that $f(\mathbf{x}^\dagger) < f(\mathbf{x}^*)$. Now, since $\mathbf{S}$ is convex, the point

$$\mathbf{x} = \lambda \mathbf{x}^* + (1 - \lambda)\mathbf{x}^\dagger \tag{B2}$$

also belongs to $\mathbf{S}, \forall \lambda \in [0, 1]$. Thus,

$$f(\mathbf{x}) = f(\lambda \mathbf{x}^* + (1-\lambda)\mathbf{x}^\dagger) \leq \lambda f(\mathbf{x}^*) + (1-\lambda)f(\mathbf{x}^\dagger) < \lambda f(\mathbf{x}^*) + (1-\lambda)f(\mathbf{x}^*)$$
$$= f(\mathbf{x}^*), \tag{B3}$$

since $f(\mathbf{x})$ is convex. Thus, since $f(\mathbf{x}) < f(\mathbf{x}^*)$ holds even as $\lambda \to 1$, where $\mathbf{x} \to \mathbf{x}^*$, $f(\mathbf{x}^*)$ cannot be a local minimum. Hence our original assumption must be false, and $f(\mathbf{x}^*)$ is therefore the global minimum.

B.1.2 Properties of the gradient

Proposition Let $f(\mathbf{x})$ be a differentiable function. The largest, negative slope of $f(\mathbf{x})$ at any given point $\mathbf{x}_0$ occurs in the direction $-\nabla f(\mathbf{x}_0)$.

Proof Let the unit vector $\mathbf{d} = (d_1, d_2, \ldots, d_n)^\mathrm{T}$ represent an arbitrary direction in $\mathbf{R}^n$. An elementary result from calculus tells us that the directional derivative $f'_\mathbf{d}(\mathbf{x}_0)$, defined as the limit

$$\lim_{t \to 0} \frac{f(\mathbf{x}_0 + t\mathbf{d}) - f(\mathbf{x}_0)}{t}, \tag{B4}$$

determines the rate of change of f in the direction $\mathbf{d}$. Now, applying the chain rule for differentiation, it is easy to show that

$$f'_\mathbf{d}(\mathbf{x}_0) = d_1 \left.\frac{\partial f}{\partial x_1}\right|_{\mathbf{x}=\mathbf{x}_0} + \cdots + d_n \left.\frac{\partial f}{\partial x_n}\right|_{\mathbf{x}=\mathbf{x}_0} = \mathbf{d}^\mathrm{T} \nabla f(\mathbf{x}_0). \tag{B5}$$

Remembering that $\mathbf{d}$ is a unit vector, it is obvious from eqn (B5) that the directional derivative takes its most negative value for

$$\mathbf{d} = -\frac{\nabla f(\mathbf{x}_0)}{|\nabla f(\mathbf{x}_0)|}, \tag{B6}$$

and the proof follows.

B.2 Genetic algorithms

B.2.1 The schema theorem

The schema theorem, first derived by Holland [32], gives an indication of the spreading of chromosome parts (known as schemata, see Section 3.2.2) associated with high fitness, in a population of a GA.

Theorem Let $\Gamma(S, g)$ denote the number of copies (in a population of N individuals) of a given schema S in generation g of a GA. The expected number of copies of S in generation $g + 1$ is then given by

$$E(\Gamma(S, g+1)) \geq \frac{\overline{F}_S}{\overline{F}} \Gamma(S, g) \left(1 - p_c \frac{d(S)}{m-1}\right)(1 - p_{\mathrm{mut}})^{o(S)} \tag{B7}$$

where $\overline{F}$ is the average fitness in the population and $\overline{F}_S$ denotes the average fitness of a schema S defined as the average fitness of those individuals whose chromosomes contain the schema in question. p_c is the crossover probability and p_{mut} is the mutation probability. $d(S)$ and $o(S)$, respectively, denote the defining length and order of the schema S (see Section 3.2.2).

Proof Assuming that the selection of individuals is carried out in proportion to their fitness, the probability of an individual with fitness F_i being selected, in a single selection step, is equal to F_i/F, where $F = \sum_{i=1}^{N} F_i$ is the total (summed) fitness in the population. The probability of selecting an individual containing S is thus equal to the ratio of the total fitness F_S contained in such individuals and the total fitness in the entire population. Since, by the definition of the arithmetic average,

$$\overline{F}(S) = \frac{F_S}{\Gamma(S,g)}, \tag{B8}$$

the probability $p_{\text{sel}}(S)$ of selecting an individual containing S in a single selection step thus equals

$$p_{\text{sel}}(S) = \frac{\overline{F}_S}{F} \Gamma(S,g). \tag{B9}$$

Now, since N independent selection steps are carried out in the formation of the new generation, the expected number of copies of S becomes

$$E(\Gamma(S,g+1)) = N \frac{\overline{F}_S}{F} \Gamma(S,g) = \frac{\overline{F}_S}{\overline{F}} \Gamma(s,g). \tag{B10}$$

In addition to selection, however, there are also the processes of crossover and mutation, which tend to destroy long schemata. Consider first crossover. We will assume that single-point crossover is used, with random selection of the crossover point. The probability of destruction, during crossover, of a schema with defining length $d(S)$ is then given by

$$p_d = \frac{d(S)}{m-1}, \tag{B11}$$

where m is the chromosome length. Thus, a schema of defining length $m-1$, i.e. one that lacks wild cards altogether, will (obviously) be destroyed with probability 1. Clearly, it is only the defining length that matters. Wild cards in the very beginning or end of a string can be replaced with any symbol (0,1 or x) without destroying the schema in question. Since crossover occurs with probability p_c, the survival probability of the schema S will be equal to $1 - p_c d(S)/(m-1)$. In fact, the probability of survival is larger than this estimate, since a schema may be reassembled during crossover. For instance, if the string 110101, which contains the schema $S_1 = 11xx01$ is cut in the middle, i.e. between the third and fourth genes, the schema is initially destroyed. However, if the first part (110) of the disassembled chromosome is reassembled with a part of the form 001 or 101, the schema still survives. Similarly, if the second half of the disassembled chromosome, 101, is reassembled with either 111 or 110, to form 111101 or 110101, respectively, the schema also survives. Hence, including the effects of crossover, the expected number of copies of S in generation $g + 1$ can be expressed as the inequality

$$E(\Gamma(S, g+1)) \geq \frac{\bar{F}_S}{\bar{F}} \Gamma(S, g) \left(1 - p_c \frac{d(S)}{m-1}\right). \tag{B12}$$

During mutation, a schema will be destroyed if any of its non–wild card genes are changed. For each gene, the probability of mutation is equal to p_{mut}. Thus, the probability that the schema S will *not* mutate in any of the fixed positions equals $(1 - p_{\text{mut}})^{o(S)}$. Thus, including the effects of both crossover and mutation, the schema theorem is obtained:

$$E(\Gamma(S, g+1)) \geq \frac{\bar{F}_S}{\bar{F}} \Gamma(S, g) \left(1 - p_c \frac{d(S)}{m-1}\right)(1 - p_{\text{mut}})^{o(S)}. \tag{B13}$$

B.2.2 The genetic algorithm as a Markov process

B.2.2.1 Number of populations of a given size

In the finite population case, the effects of finite sampling must be taken into account when studying the evolution of a population from one generation to the next. The number of possible populations of a given size grows very rapidly with the population size and the chromosome length, as the following proposition shows.

Proposition The number $v(N, m)$ of populations of size N, consisting of (binary) chromosomes of length m equals

$$v(N, m) = \binom{N + 2^m - 1}{2^m - 1}. \tag{B14}$$

Proof There are 2^m possible binary chromosomes of length m. Thus, the problem of enumerating all populations consisting of N chromosomes is equivalent to the problem of placing N balls in $k = 2^m$ containers, defined by $k - 1$ container walls. The problem is illustrated for the case $N = 4$, $m = 2$ in Fig. B.1. In order to compute $v(N, m)$ one may consider all possible placements of the $k - 1$ container walls, as indicated in the figure. Clearly, if the balls are indistinguishable, this procedure will result in all possible divisions of the N balls. The problem therefore consists of finding all possible ways of placing $k - 1$ walls among $N + k - 1$ positions. In general, the number of ways γ of selecting κ positions among K possible ones is given by

$$\gamma = \binom{K}{\kappa}. \tag{B15}$$

With $K = N + 2^m - 1$ and $\kappa = 2^m - 1$ the proposition follows.

APPENDIX B: ANALYSIS OF OPTIMIZATION ALGORITHMS 177

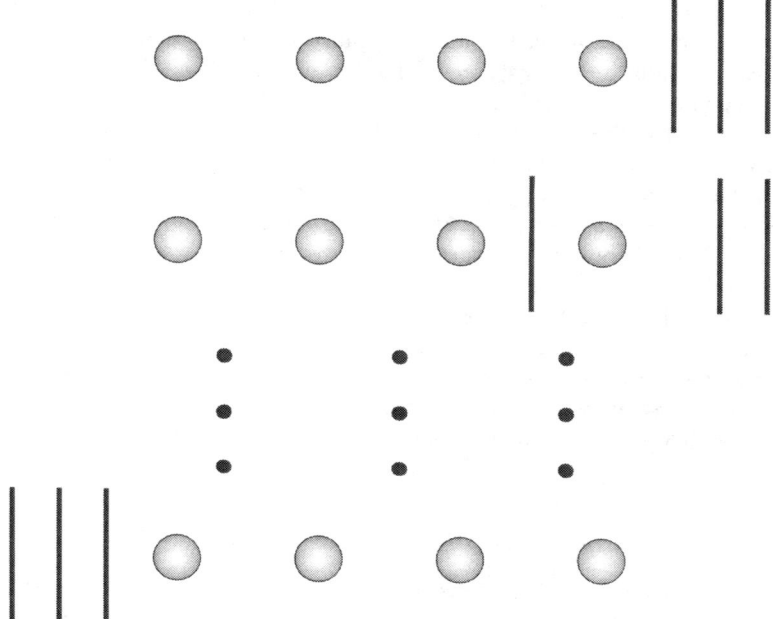

Figure B.1: *The problem of finding the number of populations of a given size N, consisting of chromosomes of length* m, *can be reduced to the problem of placing N balls in* $k-1$ *containers, where* $k = 2^m$. *The figure shows three of the 35 possible configurations for* $N = 4, m = 2$.

B.2.3 Infinite population models

B.2.3.1 Representing the crossover operator

An analytical representation of the crossover operator must take into account that, in general, not all chromosomes can be formed through crossover of two given strings. This fact can be neatly represented using the bitwise XOR ($\oplus$) and AND ($\otimes$) operators. Bitwise XOR of two bitstrings c_k and c_l is obtained by considering each position along the two strings, and setting the resulting bit to 1 if exactly one of the corresponding bits (from either string) is equal to 1. Thus, for example, with $c_k = 0101$ and $c_l = 1100$, one obtains $c_k \oplus c_l = 0101 \oplus 1100 = 1001$. Bitwise AND is computed in a similar way, but here the resulting bit is instead set to 1 only if both of the corresponding bits in c_k and c_l are equal to 1. Thus $c_k \otimes c_l = 0101 \otimes 1100 = 0100$.

Proposition The string c_j can be obtained from crossover of two strings c_k and c_l only if $(c_k \oplus c_j) \otimes (c_l \oplus c_j) = 0_m$, where 0_m denotes a chromosome consisting of m 0s.

Proof In order for a chromosome c_j to be a possible result of crossover of two strings c_k and c_l, the alleles (0 or 1) of c_j must, for all positions along the chromosome, be available in at least one of the chromosomes c_k and c_l. The operations $c_k \oplus c_j$ and $c_l \oplus c_j$ will result in strings, here called σ_{kj} and σ_{lj}, respectively, such that σ_{kj} has a 1 in those positions in which c_k and c_j differ whereas σ_{lj} has a 1 in those positions in which c_l and c_j differ. Thus, the formation of a string c_j is possible only if, for all positions along the chromosome, the bits taken from σ_{kj} and σ_{lj} are not both 1. If this condition holds, $\sigma_{kj} \otimes \sigma_{lj} = 0_m$, since the bitwise AND operator only returns a 1 if both bits are equal to 1.

B.2.3.2 Initial distribution of chromosomes

Proposition Consider an infinite set of binary chromosomes of length m. Let j denote the number of 1s in a chromosome, and let $p_1(j)$ denote the initial distribution of chromosomes. Provided that the chromosomes are generated randomly, with equal probability of setting a given gene to either 0 or 1, $p_1(j)$ satisfies

$$p_1(j) = 2^{-m} \binom{m}{j}. \tag{B16}$$

Proof The number of ways of distributing j 1s among m positions equals $\binom{m}{j}$. Since the total number of chromosomes of length m is equal to 2^m, the proposition follows.

B.2.3.3 Elementary properties of binomial coefficients

Proposition The binomial coefficients $\binom{m}{j}$ have the following properties:

$$\sum_{j=0}^{m} j \binom{m}{j} = m 2^{m-1}, \tag{B17}$$

and

$$\sum_{j=0}^{m} j^2 \binom{m}{j} = m(m+1) 2^{m-2}. \tag{B18}$$

Proof Starting from the binomial theorem

$$(a+b)^m = \sum_{j=0}^{m} \binom{m}{j} a^j b^{m-j}, \tag{B19}$$

and setting $a = x$ and $b = 1$, one obtains

$$(x+1)^m = \sum_{j=0}^{m} \binom{m}{j} x^j. \tag{B20}$$

Taking the derivative with respect to x, and multiplying the result by x, one obtains

$$xm(x+1)^{m-1} = \sum_{j=0}^{m} \binom{m}{j} j x^j. \tag{B21}$$

Setting $x = 1$, the first part of the proposition follows. If, instead, the derivative of eqn (B21) is taken, and the result is again multiplied by x, one obtains

$$xm(x+1)^{m-1} + x^2 m(m-1)(x+1)^{m-2} = \sum_{j=0}^{m} \binom{m}{j} j^2 x^j, \tag{B22}$$

so that, setting $x = 1$,

$$\sum_{j=0}^{m} \binom{m}{j} j^2 = m 2^{m-1} + m(m-1) 2^{m-2} = m(m+1) 2^{m-2}. \tag{B23}$$

B.2.3.4 The mutation operator for functions of unitation

Let $M(j, k)$ denote the probability of forming a chromosome of length m containing j 1s, through mutations applied to a chromosome with k 1s. $M(j, k)$ then satisfies [7]

$$M(j,k) = \sum_{I} \binom{k}{m_1} \binom{m-k}{m_0} p_{\text{mut}}^{m_1+m_0} (1-p_{\text{mut}})^{m-m_1-m_0}, \tag{B24}$$

where m_0 and m_1 denote the number of mutations occurring in genes taking the value (allele) 0 and 1, respectively, and I is the set

$$I = \{(m_0, m_1); m_0 - m_1 = j - k, \quad 0 \le m_0 \le m - k, \quad 0 \le m_1 \le k\}. \tag{B25}$$

Proof For a chromosome containing k ones to mutate into a chromosome with j ones, one must have

$$k + m_0 - m_1 = j. \tag{B26}$$

Clearly, m_0 and m_1 most be non-negative. Furthermore, m_0 and m_1 will be bounded from above by $m - k$ and k, respectively. Summarizing the observations one obtains the definition of the set I. Now, mutations in which m_0 0s and m_1 1s mutate occur with probability

$$\pi(m_0, m_1) = p_{\text{mut}}^{m_0 + m_1} (1 - p_{\text{mut}})^{m - m_0 - m_1}. \tag{B27}$$

Furthermore, the number of ways such mutations can occur equals

$$\nu(m_0, m_1) = \binom{k}{m_1} \binom{m-k}{m_0}. \tag{B28}$$

Thus, summarizing eqns (B25), (B27) and (B28), $M(j,k)$ is obtained as

$$M(j,k) = \sum_l v(m_0, m_1)\pi(m_0, m_1)$$
$$= \sum_l \binom{k}{m_1}\binom{m-k}{m_0} p_{\text{mut}}^{m_1+m_0}(1-p_{\text{mut}})^{m-m_1-m_0}. \quad \text{(B29)}$$

B.2.3.5 Selection and mutation for the Onemax problem

Lemma Starting from an infinite population with the distribution

$$p_1(j) = 2^{-m}\binom{m}{j}, \quad \text{(B30)}$$

where j denotes the number of 1s in a chromosome, and applying fitness-proportional selection and single-gene mutations (described in Section 3.2.2), the resulting distribution equals

$$p_2(j) = 2^{1-m}\left(\frac{j}{m} + p_\mu \frac{m-2j}{m^2}\right)\binom{m}{j}, \quad \text{(B31)}$$

where p_μ denotes the probability of mutating exactly one gene.

Proof The distribution $p_2(j)$ is obtained by first carrying out selection, then mutation. The distribution obtained after the selection step is given in eqn (3.38) and equals

$$\mathcal{G}_s(p_1)(j) = 2^{1-m}\frac{j}{m}\binom{m}{j}. \quad \text{(B32)}$$

Consider now single-gene mutations. For any given j, the contribution to $p_2(j)$ will come from three sources: chromosomes with j 1s that do not mutate, chromosomes with $j-1$ 1s in which a 0 is mutated and, finally, chromosomes with $j+1$ 1s in which a 1 is mutated. Hence

$$p_2(j) = (1-p_\mu)\mathcal{G}_s(p_1)(j) \quad \text{(B33)}$$
$$+ p_\mu\left(\frac{m-j+1}{m}\mathcal{G}_s(p_1)(j-1) + \frac{j+1}{m}\mathcal{G}_s(p_1)(j+1)\right).$$

Inserting the expression from eqn (B32), one obtains

$$p_2(j) = (1-p_\mu)2^{1-m}\frac{j}{m}\binom{m}{j}$$
$$+ p_\mu 2^{1-m}\left[\frac{(m-j+1)(j-1)}{m^2}\binom{m}{j-1} + \frac{(j+1)^2}{m^2}\binom{m}{j+1}\right]$$

$$= (1-p_\mu)2^{1-m}\frac{j}{m}\binom{m}{j} + p_\mu 2^{1-m}\frac{mj+m-2j}{m^2}\binom{m}{j}$$

$$= 2^{1-m}\left(\frac{j}{m} + p_\mu\frac{m-2j}{m^2}\right)\binom{m}{j}, \tag{B34}$$

where, in the second step, the two easily derived binomial identities

$$\binom{m}{j-1} = \frac{j}{m-j+1}\binom{m}{j} \tag{B35}$$

and

$$\binom{m}{j+1} = \frac{m-j}{j+1}\binom{m}{j} \tag{B36}$$

have been used.

B.2.4 Expected runtime of a simple GA

Lemma Consider a very simple GA, with a population consisting of a single individual that is modified through mutations only, and where a mutated individual is kept (thus replacing the parent) if the mutation is successful, i.e. if the mutated individual obtains a higher fitness than the parent. Let m denote the chromosome length, and define the mutation rate p_{mut} as k/m for some $k \ll m$. Then the expected number of evaluations, $E(L)$, needed to obtain the maximum fitness for the Onemax problem can be estimated as [52]

$$E(L) \approx e^k \frac{m}{k} \ln \frac{m}{2}. \tag{B37}$$

Proof Obviously, in this simple GA, fitness values can never decrease. In order to simplify the expressions below, the number of 0s will here be denoted l, so that the number of 1s equals $m - l$. Let $P(l, p_{mut})$ denote the probability of improving a chromosome containing l 0s. In order to improve the chromosome, more 0s than 1s must mutate. The probability for this combination of events can be approximated as

$$P(l, p_{mut}) \approx (1 - p_{mut})^{m-l}(1 - (1 - p_{mut})^l). \tag{B38}$$

This expression summarizes the case in which none of the 1s mutate, and at least one of the 0s does. Note that an exact expression would involve a sum over all possible mutations in which the terms leading to an increase in the number of 1s would contribute to $P(l, p_{mut})$. For the purposes of this approximate analysis, the expression in eqn (B38) will suffice, however. Now, given $P(l, p_{mut})$, one may estimate the expected number of evaluations $E(\Delta L(l, p_{mut}))$ needed to obtain an improvement as

$$E(\Delta L(l, p_{mut})) = \frac{1}{P(l, p_{mut})} \equiv \frac{1}{P(l, k/m)}. \tag{B39}$$

Assuming random initialization of the chromosomes, it is likely that the first individual contains around $m/2$ 0s. Thus, the expected total number of evaluations required to reach the target chromosome (with only 1s) becomes

$$E(L) = E(\Delta L(m/2, k/m)) + E(\Delta L(m/2 - 1, k/m)) + \cdots + E(\Delta L(1, k/m)). \tag{B40}$$

Using the assumption that $k \ll m$, we can write

$$P(l, k/m) = (1 - k/m)^{m-l}(1 - (1 - k/m)^l)$$

$$\approx (1 - k/m)^{m-l} lk/m \to e^{-k} lk/m \tag{B41}$$

for large values of m, so that, reversing the order of summation

$$E(L) \approx e^k \frac{m}{k} \sum_{l=1}^{m/2} \frac{1}{l}. \tag{B42}$$

Since

$$\sum_{l=1}^{m/2} \frac{1}{l} \approx \ln \frac{m}{2}, \tag{B43}$$

the lemma follows. (In fact, the difference $\sum_{l=1}^{s} 1/l - \ln s$ converges to Euler's constant $\gamma \approx 0.5772$, as $s \to \infty$).

B.2.5 Estimating optimal mutation rates

The choice of the mutation probability p_{mut} has great influence on the results obtained when running a GA, and it should therefore be chosen carefully. A common rule of thumb is to set

$$p_{\text{mut}} \approx \frac{1}{m}, \tag{B44}$$

so that, on average, around one mutation occurs per chromosome. In fact, an informal motivation for this choice can be obtained using the same simple GA as in the runtime analysis presented above.

Lemma For the simple GA with a population size of one, in which new individuals are generated through mutations and where the offspring is kept only if it is better than the parent, the optimal mutation probability, in the case of the Onemax problem, is around $1/m$.

Proof In the runtime analysis, an approximate expression for the probability of an improvement (i.e. an increase in the number of 1s in the chromosome) was given in eqn (B38). Introducing $x = 1 - p_{\text{mut}}$ that equation can be rewritten as

$$\Pi(l, x) = x^{m-l}(1 - x^l), \tag{B45}$$

where l is the number of 0s in the chromosome before any mutation takes place. The derivative of Π with respect to x is

$$\Pi'_x(l,x) = (m-l)x^{m-l-1}(1-x^l) - x^{m-l}lx^{l-1} = (m - mx^l - l)x^{m-l-1}. \quad (B46)$$

Thus, setting the derivative to zero, one finds the optimal value of x as

$$x^* = (1 - l/m)^{1/l}, \quad (B47)$$

so that the optimal mutation rate becomes

$$p^*_{\text{mut}} = 1 - (1 - l/m)^{1/l}. \quad (B48)$$

For large values of m, and for $l \ll m$, a series expansion of eqn (B48) gives

$$p^*_{\text{mut}} \approx \frac{1}{m}. \quad (B49)$$

B.3 Ant colony optimization

B.3.1 Pheromone limits in MMAS

Proposition For the Max-min ant system (MMAS) algorithm, the maximum pheromone level $\tau_{\max}$ on any edge e_{ij} is bounded (asymptotically) by f^*/ρ, where f^* is the value of the objective function for the optimal solution [70] (i.e. $1/D^*$ in the case of the TSP).

Proof Let τ_0 denote the initial amount of pheromone on the edges of the construction graph. In any given iteration, the maximum amount of pheromone that can be added to an edge e_{ij} equals f^*. Thus, after the first iteration, the maximum pheromone level equals $(1-\rho)\tau_0 + f^*$, and after the second iteration it equals $(1-\rho)^2\tau_0 + (1-\rho)f^* + f^*$. In general, after the kth iteration, the maximum amount of pheromone equals

$$\tau_{\max}(k) = (1-\rho)^k \tau_0 + \sum_{j=1}^{k}(1-\rho)^{j-1}f^*. \quad (B50)$$

Since $\rho \in \,]0,1]$, in the limit $k \to \infty$ we obtain

$$\tau_{\max} = f^*/\rho, \quad (B51)$$

using the well-known result

$$\sum_{j=1}^{\infty}(1-\rho)^{j-1} = 1/\rho. \quad (B52)$$

B.3.2 Convergence proof

Theorem Let $p(k)$ be the probability that the optimal solution (for the TSP) is encountered (using MMAS) at least once in the first k iterations. Then

$$\lim_{k\to\infty} p(k) = 1. \tag{B53}$$

Proof In MMAS, pheromone levels are bounded from below by $\tau_{\min}$, and from above by $\tau_{\max}$. In fact, the proposition above shows that, due to pheromone evaporation, an explicit upper bound is not even needed. Consider now the construction of a tour by a given ant. We can determine a lower bound for the probability[1] of traversing any edge e_{ij}, by setting the pheromone level on that edge to $\tau_{\min}$, and the pheromone level on all other edges to $\tau_{\max}$. The probability $p_{\min}$ of traversing e_{ij} is thus bounded from below by p_0, given by (see eqn (4.3))

$$p_{\min} \geq p_0 = \frac{\tau_{\min}^\alpha}{(n-1)\tau_{\max}^\alpha + \tau_{\min}^\alpha}. \tag{B54}$$

Hence, any tour (including an optimal tour S^*) will be generated with a probability p_S fulfilling

$$p_S \geq p_{\min}^n. \tag{B55}$$

Thus, the probability of at least *one* ant finding an optimal solution (assuming, for simplicity, that only one ant is evaluated in each iteration) is bounded from below by

$$p(k) = 1 - (1 - p_S)^k. \tag{B56}$$

Therefore, evidently,

$$\lim_{k\to\infty} p(k) = 1. \tag{B57}$$

B.3.3 Runtime analysis for a simple ACO algorithm

Theorem Consider the Onemax problem introduced in Chapter 3, with n bits (and, therefore, a maximum fitness of n) and an ACO algorithm using a chain construction graph (see Chapter 4), with a single ant, and with pheromone updates according to

$$\tau_{ij} \leftarrow (1 - \rho)\tau_{ij} + \rho/D_b, \tag{B58}$$

---

[1] For simplicity the problem-specific factor η_{ij} will be dropped (i.e. set to 1) in the proof. However, as long as $0 < \eta_{\min} \leq \eta_{ij} \leq \eta_{\max} < \infty, \forall i,j \in [1,n]$, the proof still holds even with η_{ij} included.

if the edge (i, j) belongs to the best path found so far, i.e. the path generating the largest number of 1s, and

$$\tau_{ij} \leftarrow (1 - \rho)\tau_{ij} \tag{B59}$$

for edges that do not belong to the best path. Nominally, D_b is the length of the best path found so far, but it is here simply taken as $1/n$, i.e. the number of non-trivial steps traversed by any ant in the chain construction graph (see Fig. B.2). Furthermore, the algorithm uses an explicit lower pheromone limit of $\tau_{\min} = a/n^2$, where a is a constant ($\ll n$) determining the evaporation rate ρ as

$$\rho = 1 - a/n, \tag{B60}$$

and initial pheromone values, on all edges, of $\tau_0 = 1/2n$. The expected runtime (number of iterations) $E(L)$ for the algorithm, applied to the Onemax problem, is then upper bounded [27] by $2e^{2a}nH_n/a$ where

$$H_n = \sum_{s=1}^{n} \frac{1}{s} \tag{B61}$$

is the nth harmonic number. For large (but finite) n, the bound is of order $n \ln n$.

Proof Even though only a lower bound on τ_{ij} is introduced, it is easy to see that an implicit upper bound of

$$\tau_{\max} = \frac{1}{n} \tag{B62}$$

holds. This is so since, for an edge that constantly belongs to the best tour, and is therefore reinforced in each step, the pheromone level τ_b after one iteration equals

$$\tau_b = (1 - \rho)\frac{1}{2n} + \frac{\rho}{n} = \frac{1}{n} - \frac{a}{2n^2}, \tag{B63}$$

and, more generally, after l iterations

$$\tau_b = \frac{1}{n} - \frac{1}{2n}\left(\frac{a}{n}\right)^l, \tag{B64}$$

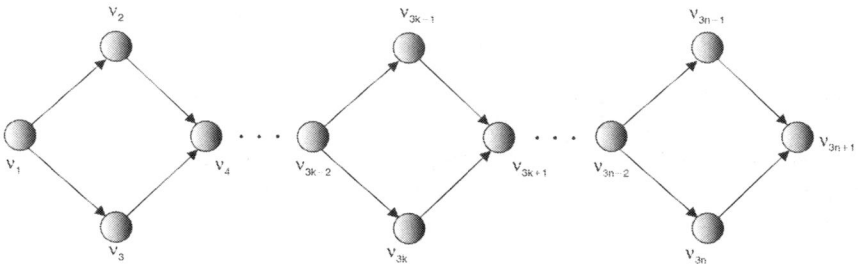

Figure B.2: *Chain construction graph used in the proof regarding runtime limits for ACO.*

which evidently converges to $1/n$ if $a < n$. Consider now the construction graph shown in Fig. B.2. Element k of the binary string of digits (of length n) is obtained by taking a step from node v_{3k-2} to either v_{3k-1} (generating a 1) or v_{3k} (generating a 0). As described in Chapter 4, the subsequent step, to node v_{3k+1} is deterministic. Now, because of the pheromone limits, the minimum and maximum probabilities of taking any non-deterministic step, generating either a 0 or a 1 (the only two options) fulfil

$$p_{\min} = \frac{\tau_{\min}}{\tau_{\min} + \tau_{\max}} = \frac{\frac{a}{n^2}}{\frac{a}{n^2} + \frac{1}{n}} = \frac{a}{n+a} \equiv \frac{q}{1+q}, \tag{B65}$$

and

$$p_{\max} = \frac{\tau_{\max}}{\tau_{\min} + \tau_{\max}} = \frac{\frac{1}{n}}{\frac{a}{n^2} + \frac{1}{n}} = \frac{n}{n+a} \equiv \frac{1}{1+q}, \tag{B66}$$

where we have introduced

$$q = 1 - \rho = \frac{a}{n}. \tag{B67}$$

In each iteration, the edges along the best path (found so far) are reinforced. The pheromone level τ_{re} on *any* edge that is reinforced satisfies

$$\tau_{re} \geq (1-\rho)\tau_{\min} + \frac{\rho}{n} = q\tau_{\min} + \frac{1-q}{n} \geq \frac{1-q}{n}. \tag{B68}$$

Similarly, for an edge that is *not* reinforced, the pheromone level τ_{nr} is bounded from above as follows

$$\tau_{nr} \leq (1-\rho)\tau_{\max} = q\tau_{\max} = \frac{q}{n} = \frac{a}{n^2} = \tau_{\min}. \tag{B69}$$

Since $\tau_{\min}$ also is an explicit lower limit for the pheromone level, one obtains, in fact

$$\tau_{nr} = \tau_{\min}. \tag{B70}$$

Now, at any iteration, the best path consists of n non-deterministic steps. Let $p_k^{[re]}$ denote the probability, given the pheromone levels, of making, in step k, the move corresponding to the best path, and, let $p_k^{[nr]}$ denote the probability of making, in step k, the move that does *not* correspond to the best path. In view of the bounds on pheromone levels just given, one obtains

$$p_k^{[nr]} = \frac{\tau_{nr}}{\tau_{nr} + \tau_{re}} \leq \frac{\frac{q}{n}}{\frac{q}{n} + \frac{1-q}{n}} = q, \tag{B71}$$

so that

$$p_k^{[re]} = 1 - p_k^{[nr]} \geq 1 - q. \tag{B72}$$

Now, let A_s denote the set of all binary strings with fitness s, i.e. strings containing s 1s. Consider the case where the current best path generates a string belonging to A_s for some value of $s \in [0, n-1]$. This path is then reinforced according to eqn (B58), and one must now determine the probability π of finding a better solution in the next iteration. Clearly, for an improvement to occur, a net sum of between one and $n-s$ 0s must be changed to 1s. One can find a lower bound for the probability π, by only considering events in which *exactly* one 0 is changed to a 1. There are $n-s$ such moves, all of which are mutually exclusive. For any such move the probability of changing a 0 to a 1 is larger than or equal to $p_{\min}$, see eqn (B65), and the probability of *not* changing any of the other $n-1$ positions is larger than or equal to $(1-q)^{n-1}$, see eqn (B72), so that

$$\pi \geq \frac{q}{1+q}(1-q)^{n-1}(n-s). \tag{B73}$$

The expected number of iterations n_s before a new, better path is found, i.e. one that belongs to $A_{s+1} \cup \ldots \cup A_n$, is thus bounded from above by

$$n_s = \frac{q+1}{q}(1-q)^{-n+1}\frac{1}{n-s}. \tag{B74}$$

In the worst case scenario, the algorithm starts with a path consisting of only 0s, and proceeds by improving one position (along the binary string) at a time, until the string consists of only 1s. In this case, the total expected number of iterations $E(L)$ satisfies

$$E(L) = \sum_{s=0}^{n-1}\frac{q+1}{q}(1-q)^{-n+1}\frac{1}{n-s} = \frac{q+1}{q}(1-q)^{-n+1}\sum_{s=1}^{n}\frac{1}{s}$$

$$= \frac{a+n}{a\left(1-\frac{a}{n}\right)^{n-1}}H_n < \frac{2}{a\left(1-\frac{a}{n}\right)^{n-1}}nH_n < \frac{2e^{2a}}{a}nH_n, \tag{B75}$$

where, in the last step, it has been noted that

$$(n-1)\ln\left(1-\frac{a}{n}\right) > (n-1)\left(-\frac{2a}{n}\right) > -2a \tag{B76}$$

for $a \leq n/2$, so that $e^{2a} > (1-a/n)^{-n+1}$. For large (but finite) n, H_n is of order $\ln n$, so that $E(L)$ is of order $n \ln n$.

B.4 Particle swarm optimization

B.4.1 Particle trajectories in PSO

Consider a simplified PSO algorithm, in which (1) the random factors r and q in eqn (5.14) have been removed (i.e. set to 1), (2) a single particle is considered, in one dimension, starting at $x = x_0$ with velocity $v = v_0$, and (3) the particle-best and swarm-best positions, x^{pb} and x^{sb} are fixed. eqn (5.14) can then be written

$$v \leftarrow v + c_1(x^{\text{pb}} - x) + c_2(x^{\text{sb}} - x). \tag{B77}$$

Similarly, eqn (5.15) can be written

$$x \leftarrow x + v \tag{B78}$$

where, in both equations, Δt has been set to 1.

Lemma For the simple PSO just introduced, the particle trajectories remain bounded only if $c_1 + c_2 < 4$.

Proof Let k denote the kth iteration of the algorithm, so that

$$v(k+1) = v(k) + c_1(x^{\text{pb}} - x(k)) + c_2(x^{\text{sb}} - x(k)), \tag{B79}$$

and

$$x(k+1) = x(k) + v(k+1). \tag{B80}$$

Defining

$$c = c_1 + c_2 \tag{B81}$$

and

$$x^b = \frac{c_1 x^{\text{pb}} + c_2 x^{\text{sb}}}{c_1 + c_2}, \tag{B82}$$

eqn (B79) can be rewritten as

$$v(k+1) = v(k) + c(x^b - x(k)). \tag{B83}$$

Introducing $y(k) = x^b - x(k)$, eqns (B83) and (B80) can be written as

$$v(k+1) = v(k) + cy(k), \tag{B84}$$

and

$$y(k+1) = y(k) - v(k+1) = -v(k) + (1-c)y(k). \tag{B85}$$

Introducing the vector $\mathbf{s}(k) = (v(k), y(k))^T$, eqns (B84) and (B85) can be written in matrix form as [9]

$$\mathbf{s}(k+1) = M\mathbf{s}(k), \tag{B86}$$

where

$$M = \begin{pmatrix} 1 & c \\ -1 & 1-c \end{pmatrix}. \tag{B87}$$

Thus, in general,

$$\mathbf{s}(k) = M^k \mathbf{s}(0). \tag{B88}$$

The movement of the particle is thus fully determined by the matrix M. In order to simplify the analysis, we should diagonalize M, i.e. find a matrix A such that

$$L = A^{-1}MA \tag{B89}$$

is diagonal. The elements of L will then be the eigenvalues of M, i.e.

$$L = \begin{pmatrix} e_1 & 0 \\ 0 & e_2 \end{pmatrix}. \tag{B90}$$

If such an invertible matrix can be found, one may introduce

$$\mathbf{q}(k) = A^{-1}\mathbf{s}(k), \tag{B91}$$

and rewrite eqn (B86) as

$$\mathbf{q}(k+1) = A^{-1}MA\mathbf{q}(k) = L\mathbf{q}(k), \tag{B92}$$

so that

$$\mathbf{q}(k) = L^k \mathbf{q}(0) = \begin{pmatrix} e_1^k & 0 \\ 0 & e_2^k \end{pmatrix} \mathbf{q}(0). \tag{B93}$$

It turns out (check!) that the matrix

$$A = \begin{pmatrix} -\frac{c}{2} - \frac{\sqrt{c^2-4c}}{2} & -\frac{c}{2} + \frac{\sqrt{c^2-4c}}{2} \\ 1 & 1 \end{pmatrix} \tag{B94}$$

diagonalizes M for $c \neq 4$. If $c = 4$, A is not invertible. We shall first consider the case $c < 4$. The eigenvalues e_1 and e_2 of the matrix M can easily be determined (cf. eqn (2.11)) as

$$e_{1,2} = 1 - \frac{c}{2} \pm \frac{\sqrt{c^2-4c}}{2}. \tag{B95}$$

Now, introducing the notation $\mathbf{q}(k) = (q_1(k), q_2(k))^T$, it is easy to see that $q_1(k) = e_1^k q_1(0)$ and $q_2(k) = e_2^k q_2(0)$, and the boundedness, or absence thereof, will thus depend on the moduli of the eigenvalues e_1 and e_2. For $c < 4$, e_1 and e_2 are, in fact, complex,[2] and can be written

$$e_{1,2} = 1 - \frac{c}{2} \pm i \frac{\sqrt{4c - c^2}}{2}, \tag{B96}$$

with moduli

$$|e_1| = |e_2| = \left(1 - \frac{c}{2}\right)^2 + \frac{4c - c^2}{4} = 1. \tag{B97}$$

Thus, $|q_1(k)| = |q_1(0)|$ and $|q_2(k)| = |q_2(0)|$, so that $\mathbf{q}$ and, therefore, $\mathbf{s}$ both remain bounded as k tends to infinity, if $c = c_1 + c_2 < 4$.

In order to prove that the trajectory diverges for $c_1 + c_2 \geq 4$, we shall use a different technique [54]: combining eqn (B80) and eqn (B83) one easily obtains

$$x(k+2) + (c-2)x(k+1) + x(k) = cx^b, \tag{B98}$$

with the initial conditions $x(0) = x_0$ and $x(1) = v_0 + cx^b + x_0(1-c)$. eqn (B98) is a second order difference equation, the solution of which can be written as

$$x(k) = x^{\text{hom}}(k) + x^{\text{part}}(k), \tag{B99}$$

where $x^{\text{part}}(k)$ is any solution to the difference equation, and $x^{\text{hom}}(k)$ is the solution to the homogeneous equation

$$x(k+2) + (c-2)x(k+1) + x(k) = 0. \tag{B100}$$

Starting with $x^{\text{part}}(k)$, it is easy to see that $x(k) = x^b$ satisfies eqn (B98). The solution to a homogeneous difference equation of the form

$$x(k+2) + ax(k+1) + bx(k) = 0, \tag{B101}$$

is given by[3]

$$x^{\text{hom}}(k) = \alpha r_1^k + \beta r_2^k, \tag{B102}$$

where α and β are constants and $r_{1,2}$ are the solutions to the characteristic equation

$$r^2 + ar + b = 0. \tag{B103}$$

In this case, $a = c - 2$ and $b = 1$, so that

$$r_{1,2} = 1 - \frac{c}{2} \pm \frac{\sqrt{c^2 - 4c}}{2}. \tag{B104}$$

---

[2] The vector $\mathbf{s}(k)$, obtained as $A\mathbf{q}(k)$, is of course real-valued.
[3] The form given in eqn (B102) applies to the case $r_1 \neq r_2$. If $r_1 = r_2$, the solution instead takes the form $x^{\text{hom}}(k) = (\alpha k + \beta) r_1^k$.

Determining the coefficients α and β to match the initial conditions given above, one finds after some algebra

$$\beta = \frac{(x_0 - x^b)(\sqrt{c^2 - 4c} + c)}{2\sqrt{c^2 - 4c}} - \frac{v_0}{\sqrt{c^2 - 4c}}, \qquad \text{(B105)}$$

and

$$\alpha = x_0 - x^b - \beta. \qquad \text{(B106)}$$

Now, if $c > 4$, $r_{1,2}$ will be real-valued. Moreover, $|r_2| > 1$, so that $x(k)$ increases without bound (provided that $\beta \neq 0$). For the special case $c = 4$, one can easily show that $|x(k)|$ increases linearly with k. Thus, the trajectories remain bounded only for $c = c_1 + c_2 < 4$.

Appendix C: Data analysis

This appendix deals with data analysis in a general sense, covering two different topics: (1) hypothesis evaluation, which can, for example, be used in performance comparison of different stochastic optimization algorithms, and (2) experiment design in optimization problems involving a fixed data set of limited size. These are important issues in their own right, but here we shall consider only those aspects that are relevant to the topics covered in this book.

C.1 Hypothesis evaluation

By their very nature, stochastic optimization algorithms will generate different results when run repeatedly. As we noted in Section 1.2, this is not really a weakness and, furthermore, the stochastic nature of such algorithms allows them to avoid being trapped in local optima, thus increasing the likelihood of finding the global optimum in complex optimization problems. However, it does complicate the comparison of different stochastic algorithms, as well as the comparison with deterministic optimization algorithms such as gradient descent (see Chapter 2). Consider a situation where two different stochastic optimization algorithms have been run K times. To be more specific, we may consider two instances of a GA, each with a specific parameter setting (i.e. crossover rate, mutation rate, etc.). Running each algorithm for a given number of evaluations, two comparable distributions of results (i.e. fitness values) are obtained for the two algorithms. Comparing such distributions, in order to determine whether they are truly different, i.e., whether one algorithm really performs better than the other, is the topic of **hypothesis evaluation**.

Consider a stochastic optimization algorithm with given parameter settings. In the case of a standard GA (see Algorithm 3.2), the parameters would be the population size, the selection method (and the parameters thereof, if any), the crossover probability, and the mutation rate. Finally, one must also specify the number of evaluated *individuals*[1] in each run.

Now, running an algorithm repeatedly is equivalent to sampling the distribution of results, for the parameter settings in question, and it is such (finite) sets of samples that must form the basis for the performance analysis.

Consider first the general case of a random variable X, taking values in a subset $\Omega \subseteq \mathbf{R}$. The **probability distribution** of X, denoted $p(X)$ determines the probability of finding a values in a given range when sampling X. Specifically, the probability of obtaining a sample in the range $[a, b]$ is given by

$$\text{prob}(X \in [a, b]) = \int_a^b p(X) \mathrm{d}X. \tag{C1}$$

All probability distributions satisfy the condition[2]

$$\int_\Omega p(X) \mathrm{d}X = 1. \tag{C2}$$

In general, the true mean of a random variable X is defined as the **expectation value**, i.e.

$$\mu_X = E(X) = \int_\Omega X p(X) \mathrm{d}X. \tag{C3}$$

In the (common) case where the underlying distribution $p(X)$ is unknown, given K samples of X with values X_i, $i = 1, \ldots, K$, the true mean can be estimated as average (the arithmetic mean) of the samples

$$\mu_X \approx \overline{X} = \frac{1}{K} \sum_{i=1}^{K} X_i. \tag{C4}$$

The sample average is sometimes written $\overline{X}_K$, to indicate the number of samples used when forming the average. In addition to the sample average, a measure of variation (around the mean value) is also commonly computed. Formally, the **variance** of f is defined as

$$\sigma^2 = E((X - \mu_X)^2) = \int_\Omega (X - \mu_X)^2 p(X) \mathrm{d}X. \tag{C5}$$

---

[1] A fair comparison of different stochastic optimization algorithms should be based on the number of evaluations of the objective function (i.e. the number of evaluated individuals, in the case of GAs) rather than the number of iterations (generations, in the case of GAs).

[2] If the random variable takes discrete values as, for example, in Appendix B, Section B.2.3, the integral is replaced by a sum.

The square root σ of the variance is referred to as the **standard deviation**. Naively, one may take the expression

$$s_K^2 = \frac{1}{K} \sum_{i=1}^{K} (X_i - \overline{X})^2 \tag{C6}$$

as an estimate of the variance. However, this expression contains the estimate $\overline{X}$ rather than the (unknown) true sample mean μ_X. In fact, we can rewrite the expression for s_K^2 as

$$s_K^2 = \frac{1}{K} \sum_{i=1}^{K} (X_i - \overline{X})^2 = \frac{1}{K} \left[\sum_{i=1}^{K} (X_i - \mu_X)^2 - K(\mu_X - \overline{X})^2 \right], \tag{C7}$$

so that, making use of eqn (C4),

$$E\left(s_K^2\right) = \frac{1}{K} E\left(\sum_{i=1}^{K} (X_i - \mu_X)^2\right) - E((\mu_X - \overline{X})^2) = \frac{K-1}{K} \sigma^2. \tag{C8}$$

Thus, in order to obtain an unbiased estimate of the true variance, one should instead use

$$s^2 = \frac{1}{K-1} \sum_{i=1}^{K} (X_i - \overline{X})^2. \tag{C9}$$

One of the most commonly occurring probability distributions is the **normal distribution**, given by

$$p(X) = \frac{1}{\sigma\sqrt{2\pi}} e^{-\frac{(X-\mu)^2}{2\sigma^2}}. \tag{C10}$$

In general, a random number X that follows a normal distribution with mean μ and standard deviation σ is denoted $X \sim N(\mu, \sigma^2)$. The importance of this distribution can be understood by considering the **central limit theorem**, which says that, *regardless* of the underlying (true) distribution of a random variable X, as long as it has finite mean μ and finite variance $\sigma > 0$, the distribution of the sample average $\overline{X}_K$ tends to a normal distribution as K tends to infinity. One can show that the distribution of $\overline{X}_K$, referred to as the **sampling distribution of the mean**, satisfies

$$\frac{\overline{X}_K - \mu}{\frac{\sigma}{\sqrt{K}}} \sim N(0, 1), \tag{C11}$$

as $K \to \infty$. Note that the *mean* of the sampling distribution of the mean will always equal the mean of X. The standard deviation of the sampling distribution, $\sigma/\sqrt{K}$, is called the **standard error** of the mean.

When comparing stochastic optimization algorithms, the distribution of results certainly does not always follow a normal distribution. For example, when minimizing a function $f(\mathbf{x})$, with a minimum value $f(\mathbf{x}^*) = 0$, one typically finds that most runs generate a best value around 0 (and, obviously, no values below 0), but with a few outliers at much larger values. Thus, even though the results themselves may not be normal distributed, one may consider the sampling distribution of the mean, which will be asymptotically normal distributed, by running the algorithm K_{tot} times, and then forming K estimates of the mean, using $K_m = K_{\text{tot}}/K$ samples to form each estimate. If the true distribution does not severely deviate from a normal distribution, around $K = 30$ samples of the mean is normally sufficient for the distribution of $\overline{X}_K$ to approximate a normal distribution very well.

Example C.1
The left panel of Fig. C.1 shows as probability distribution $p(X)$ that is distinctly different from a normal distribution. In order to investigate the distribution of the *mean* $\overline{X}$ of samples taken from the distribution $p(X)$, the distribution was sampled 100,000,000 times, and the samples were collected into 100,000 groups of 1,000 samples each. Next, the average $\overline{X}_{1000}$ was formed for each group. The right panel of Fig. C.1 shows a histogram of the resulting distribution. As the figure indicates, despite the shape of $p(X)$, the average $\overline{X}$ follows a normal distribution, centered on the true mean $\mu_X = 0.46$. ∎

As an aside, in view of the omnipresence of normal distributions, knowing how to generate random numbers drawn from a normal distribution is of some importance. One method for doing so is the **Box–Müller transform**. Let r_1 and r_2 be two random variables drawn from a uniform distribution in [0, 1]. Then the variables

$$X_0 = \sqrt{-2 \ln r_1} \cos(2\pi r_2), \tag{C12}$$

and

$$X_1 = \sqrt{-2 \ln r_1} \sin(2\pi r_2) \tag{C13}$$

will be normal distributed with mean 0 and standard deviation (σ) 1, i.e. $X_1, X_2 \sim N(0, 1)$. Note also that if $X \sim N(0, 1)$, then $aX + b \sim N(b, a^2)$.

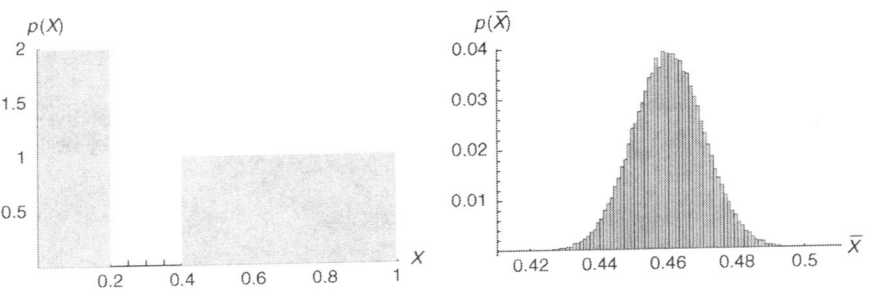

Figure C.1: *Left panel: the probability distribution $p(X)$ considered in example C.1. Right panel: the distribution of the sample average $\overline{X}$ obtained from $p(X)$.*

Returning to hypothesis evaluation, the first step of such a test consists of formulating a **null hypothesis** H_0. For example, when comparing two algorithms, the null hypothesis might be formulated as

$$H_0 : \mu_1 = \mu_2, \qquad (C14)$$

where μ_1 and μ_2 denote the mean performance of the first and second algorithm, respectively. Thus, the null hypothesis states that the two algorithms perform equally well, on average.

The null hypothesis is *rejected* if one can say with a certain degree of confidence, for example 95% or 99% that it is false. It is important to remember that if the hypothesis cannot be rejected, at a given level of confidence, this is *not* the same as having found evidence *in favour* of the hypothesis.

The procedure for investigating whether or not the null hypothesis can be rejected rests on the concept of **confidence intervals**.[3] If a measured value of a parameter falls outside a confidence interval given *a priori*, the null hypothesis should be rejected.

Here, we shall mainly concern ourselves with the problem of comparing two distributions, with the explicit purpose of determining whether one algorithm (with given parameter settings) outperforms another when applied to a particular optimization problem. However, let us begin with a simple example involving a single distribution.

Example C.2
Consider the problem of determining whether a coin is balanced, i.e. whether tossing it will generate heads and tails with equal probability. Let p_h denote the probability of the coin landing on a given side (heads, say). The null hypothesis will thus be

$$H_0 : p_h = \frac{1}{2}. \qquad (C15)$$

Regardless of the value of p_h, the number (k) of heads obtained after n tosses of the coin will then follow the **binomial distribution**.

$$p(k;n) = \binom{n}{k} p_h^k (1 - p_h)^{n-k}. \qquad (C16)$$

If n is large, the binomial distribution can be approximated well by a normal distribution with $\mu = np_h$ and $\sigma^2 = np_h(1 - p_h)$. Assuming that the coin is thrown, say, 1,000 times, then under the null hypothesis we have $\mu = 500$ and $\sigma^2 = 250$.

The distribution of k (the number of heads), is shown in Fig. C.2. As can be seen in Table C.1, for a normal distribution k will be located, with 95% probability, within the interval $[\mu - 1.96\sigma, \mu + 1.96\sigma]$. Thus, before carrying out the 1,000 tosses of the coin, we may decide to reject the null hypothesis if the number of heads lies outside the interval $[\mu - 1.96\sigma, \mu + 1.96\sigma] = [469, 531]$. If this happens, the null hypothesis can be said to

---

[3] An x % confidence interval $[a_L, a_U]$ for a parameter c, is an interval which contains c with x % probability.

198 BIOLOGICALLY INSPIRED OPTIMIZATION METHODS

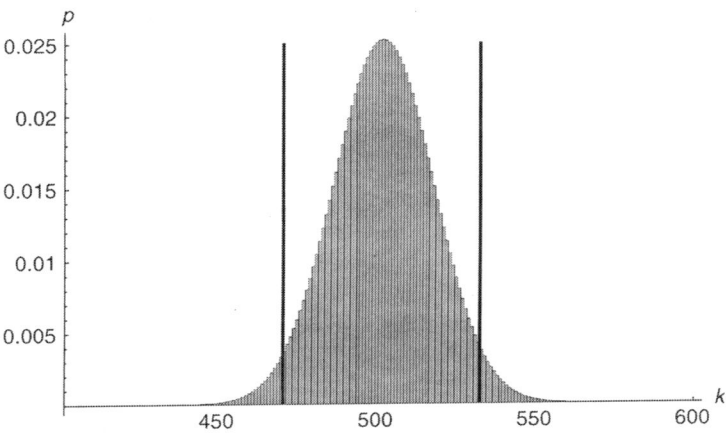

Figure C.2: *The distribution p of the number of heads obtained when tossing a balanced coin 1,000 times. The vertical lines define the 95% confidence interval for the number of heads k.*

Table C.1: *The table lists values of α such that a sample of a random variable $X \sim N(\mu, \sigma^2)$ will lie in the interval $[\mu - \alpha\sigma, \mu + \alpha\sigma]$ with probability p_α.*

p_α	α
0.90	1.645
0.95	1.960
0.98	2.326
0.99	2.576
0.999	3.291

be rejected at a **p-value** of 0.05 (i.e. $1 - 0.95$). If instead a p-value of 0.01 is required, we would reject the null hypothesis only if the number of heads lies outside the interval $[\mu - 2.58\sigma, \mu + 2.58\sigma] = [459, 541]$. See also Table C.1. ∎

Let us now proceed to the case of comparing two normal distributions, assuming that the first distribution has been sampled n_1 times and the second n_2 times. If the variances σ_1^2 and σ_2^2 of the two distributions are known, letting $\overline{X}^{[1]}$ and $\overline{X}^{[2]}$ denote the sample means of the two distributions, and μ_1 and μ_2 the true means, the quantity

$$Z = \frac{(\overline{X}^{[1]} - \overline{X}^{[2]}) - (\mu_1 - \mu_2)}{\sqrt{\sigma_1^2/n_1 + \sigma_2^2/n_2}}, \qquad (C17)$$

is normal distributed, with mean 0 and variance 1, i.e. $Z \sim N(0, 1)$, and a procedure similar to the one in Example C.2 can be applied. However, in the applications considered here, little is usually known *a priori* regarding the distributions under study. For example, the underlying variance is rarely known. In such cases, one must estimate the variance and this estimate, in turn, will follow a certain distribution (which will not be given here). One can show [20] that, if the variances of the (normal) distributions are equal, but unknown, i.e. $\sigma_1 = \sigma_2 = \sigma$, letting s_1^2 and s_2^2 denote the unbiased estimates of the standard deviation according to eqn (C9), the statistic T given by

$$T = \frac{(\overline{X}^{[1]} - \overline{X}^{[2]}) - (\mu_1 - \mu_2)}{S\sqrt{\frac{1}{n_1} + \frac{1}{n_2}}}, \tag{C18}$$

where

$$S^2 = \frac{(n_1 - 1)s_1^2 + (n_2 - 1)s_2^2}{n_1 + n_2 - 2}, \tag{C19}$$

follows a **Student's t-distribution** [24] with $n_1 + n_2 - 2$ degrees of freedom (DOFs). The statement that a random variable X follows this distribution, with ν degrees of freedom, is abbreviated $X \sim T_\nu$. The distribution is tabulated in Table C.2 for various values of the number of DOFs. Note that, for large values of the number of DOFs, the t-distribution converges to the normal distribution.

The case in which it is not even known whether the variances of the two distributions are equal is, in fact, quite difficult to treat mathematically [66]. A common choice is still to use the T-statistic, but with the number of DOFs D given by

$$D = \frac{(n_1 - 1)(n_2 - 1)}{(n_2 - 1)\delta^2 + (n_1 - 1)(1 - \delta^2)}, \tag{C20}$$

where

$$\delta = \frac{\frac{s_1^2}{n_1}}{\frac{s_1^2}{n_1} + \frac{s_2^2}{n_2}}. \tag{C21}$$

Table C.2: *The table lists the values of α, such that a sample of a random variable $X \sim T_\nu$ will lie in the interval $[-\alpha, \alpha]$, with probability p_α, for three different values of ν.*

p_α	$\nu = 50$	$\nu = 100$	$\nu = 1,000$
0.90	1.676	1.660	1.646
0.95	2.009	1.984	1.962
0.98	2.403	2.364	2.330
0.99	2.678	2.626	2.581
0.999	3.496	3.390	3.300

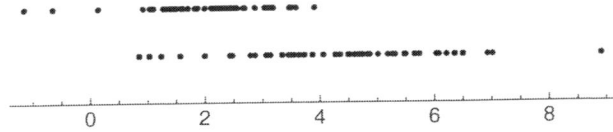

Figure C.3: *Two sets generated by sampling two random variables $X^{[1]}$ (lower row of points) and $X^{[2]}$ (upper row). See Example C.3.*

In this case, the number of DOFs will not necessarily be an integer, and must therefore be rounded off to the nearest integer. As expected, the tests using the T-statistic described above become more reliable as the number of samples increases. As a rule of thumb, the tests should be applied only if $n_1 + n_2$ is of order 30 or larger.

Example C.3
Two sets of samples taken from normal distributions with unknown mean and variance are given. The samples are displayed in Fig. C.3. Can the null hypothesis $\mu_1 = \mu_2$ be rejected at a p-value of 0.001?

Solution The sample mean and variance were computed for each of the two given sets, resulting in $\overline{X}^{[1]} = 4.38$, $s_1 = 1.65$, $\overline{X}^{[2]} = 2.04$ and $s_2 = 1.01$. There is no reason to assume that the two distributions have equal variance. Thus, the number of DOFs should be determined from eqn (C20) resulting, in this case, in a (truncated) value of $D = 81$. The T-statistic can also be computed, giving $T = 3.81$. Consulting Table C.2, we note that the null hypothesis can indeed be rejected, at a p-value of 0.001, the limiting value for such a rejection being somewhere in the range [3.39, 3.50] for a T-distribution with 81 DOFs. As an aside, note that the sets of samples were generated using the distributions $X^{[1]} \sim N(4, 4)$ and $X^{[2]} \sim N(2, 1)$. ∎

C.2 Experiment design

A common situation in many (though not all) problems involving stochastic optimization is that one has, at one's disposal, a fixed and finite set of (noisy) data, that are to be used when optimizing a computational structure with respect to some given criterion. To be specific, a common example is function fitting, where one must use a finite set of function samples in order to optimize, for example, a neural network so that it can represent the function over its entire domain, or some subset thereof. Thus, in this example, the computational structure is a neural network, and the criterion amounts to minimization of the difference between the actual function values (samples) and the output from the neural network.

Another important example is data classification where, typically, a data matrix is given, such that the columns represent different samples (patients, say, in the case of a medical application), and the rows represent attributes of the samples (e.g. age, weight, smoking habits, etc.). In addition, a class label is associated with each sample. In the special case of binary classification, there are only two classes (e.g. healthy and ill). Here, the objective is to optimize a computational structure (which might be a neural network or some other structure, see Section 3.6.3) such that, given the attributes of a sample as input, the output of the system would equal the class to which that sample belongs.

A third example is time series prediction, where the goal is to generate a system that outputs an accurate approximation of the next data point in a time series, given earlier data points as inputs.

In all these cases, the aim is to find a system (the optimized computational structure) capable of generalizing, i.e. providing correct output even for previously unseen input–output pairs. However, if all the available data are used during optimization, it is common to find (should one later obtain additional data) that the optimized system does not generalize very well at all. It is easy to see why this would be the case: consider, for example, the optimization of FFNNs with a single hidden layer of neurons (see Appendix A, Section A.2) in the case of a classification task. The parameters of the network are the weights and biases and, by allowing a large number of hidden units in the network, one may increase the number of tunable parameters to any value, however large. Now, given enough training, for example in the form of very long runs using backpropagation (see Appendix A, Section A.2.2) or PSO (see Chapter 5), a large neural network can be made to fit almost any (finite) data set, but the predictive power will drop, as the training tunes the network even to the *noise* in the data. This phenomenon, which certainly is not limited to neural networks, is known as **overfitting**.

There are various methods for avoiding overfitting. Here we shall consider only one, namely **holdout validation**. The procedure is very simple: given a sufficiently large data set,[4] the available data points are divided into three sets: a **training set**, a **validation set**, and a **test set**.[5] For ease of presentation, we shall now consider the specific case of (feedforward) neural network optimization using backpropagation, even though exactly the same principles would apply also if one instead were to use any other computational structure, and any other stochastic optimization algorithm such as, for example, PSO.

The training set is used during optimization, i.e. the optimization algorithm is given information concerning the performance of the network over this data set. In the case of backpropagation, a run though the input–output pairs in the training set constitutes one training epoch, and the RMS error over the training set is an appropriate measure of the network's overall performance, measured at the end of a training epoch.

---

[4] An exact definition of what constitutes a *sufficiently large* data set is hard to give but, as a rule of thumb, the set should contain several samples per fitted parameter.

[5] The division can clearly be done in many different ways, and there are no exact rules for optimal division of data sets. However, as a rule of thumb, it is common to divide the available data such that around 70% is used for training, around 20% for validation, and the remaining 10% for testing.

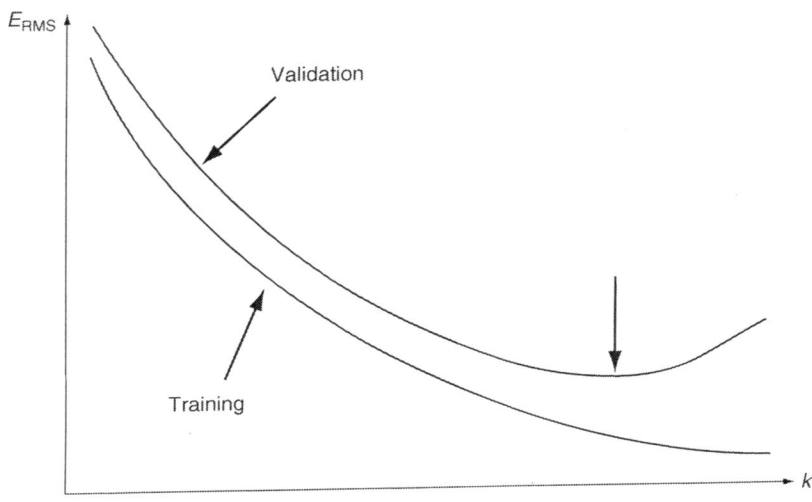

Figure C.4: *An illustration of overfitting. The horizontal axis measures the number of training epochs k. As the training progresses, the training error E_{RMS}^{tr} is gradually reduced. The validation error E_{RMS}^{val} also drops, in the beginning. However, in later epochs, the validation error begins to rise, as the network starts fitting the noise in the data. The best network is taken as the one having minimal validation error, as indicated by a vertical arrow in the figure. Note that the figure is idealized: in reality, both the training and validation errors typically display a certain amount of noise, making it more difficult to pinpoint the best network.*

The validation set, by contrast, is used for determining when to *stop* the training in order to avoid overfitting. Thus, the performance of the neural network over the validation set is *measured*, but no feedback is given to the optimization algorithm regarding its performance over the samples in the validation set. If such feedback were to be given, the validation set would, per definition, simply be a part of the training set! Now, plotting the performance of the neural network over the training and validation sets, one typically (though not always) obtains curves of the kind shown in Fig. C.4. From the figure, it is evident that, even though the error E_{RMS}^{tr} over the training set keeps falling as the training proceeds, the error E_{RMS}^{val} over the validation set reaches a minimum and then begins to rise again, indicating that overfitting is taking place. Thus, based on the information contained in the two curves shown in Fig. C.4, the best network, i.e the one with the highest ability to generalize, would be the one corresponding to minimal *validation* error. Obviously, it is not known *a priori* where this minimum will occur. During training one must therefore store the network with the current lowest validation error.

The test set is used for final evaluation of the best network obtained. One may think that the validation set would be sufficient for this purpose, but it is important to keep in mind that the network is explicitly chosen as the one generating minimal

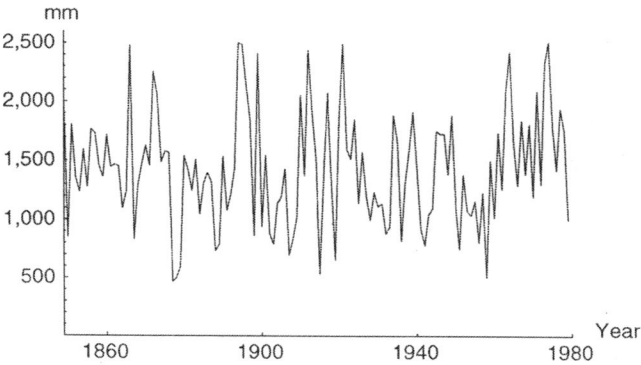

Figure C.5: *The annual rainfall in Fortaleza, Brazil, from 1849 to 1979. See Example C.4.*

error over the validation set that, of course, is of limited size. Thus, there is still a possibility (however remote, if the data sets are sufficiently large) that the good result obtained on the validation set is more due to chance than an actual ability to generalize. Of course, even with a test set, that possibility can never be completely eliminated, but if the network performs well on the test set, it is likely that it will do well also on any new data that might appear. To conclude this appendix, let us consider a specific example.

Example C.4

As an example of overfitting, we shall consider a meteorological time series, with data concerning the annual rainfall in Fortaleza, Brazil, measured between 1849 and 1979 (131 measurements, in total) [31]. The time series $x(k)$, $k = 1, \ldots, 131$ is shown in Fig. C.5. An FFNN was used for generating predictions. The time series was first converted to a form suitable for use with an FFNN, by using the elements $x(k-1), x(k-2), x(k-3), x(k-4)$ and $x(k-5)$ of the time series to form the prediction $\hat{x}(k)$ of $x(k)$, where $k = 6, \ldots, 95$ for the training set and $k = 101, \ldots, 131$ for the validation set. The indices were chosen so as to avoid any overlap between the two sets. Thus, for the training set, a total of 90 input–output pairs were generated, whereas the validation set consisted of 31 input–output pairs. Since there were only rather few data points, and since the aim of the experiment is only to demonstrate overfitting, no separate test set was generated in this case.

The FFNN had five inputs, three hidden neurons, and one output neuron. The logistic sigmoid, with $c = 2.5$, was used as the activation function. Since the measured values of annual rainfall ranged from around 400 mm to around 2600 mm, the data had to be rescaled to the range [0, 1] by first subtracting 400 from each element of the input–output pairs and then dividing the result by 2200.

Once the data sets had been generated, the FFNN was trained using backpropagation, with $\eta = 0.03$. The resulting RMS errors over the training and validation sets are shown in Fig. C.6. As is evident from the figure, the training error falls off nicely, from an initial value of around 0.2193 to around 0.1673 at the end of the run. Yet, the validation error starts with a rapid drop from around 0.2304 to 0.2113 after 1750 training epochs, before it starts rising again, eventually reaching values above 0.29. Evidently, this is a clear case of overfitting.

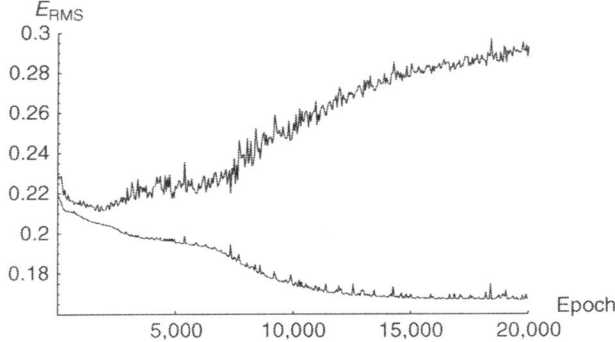

Figure C.6: *An example of overfitting. The lower curve shows the training error for an FFNN being trained to predict the rainfall data in Example C.4, whereas the upper curve shows the validation error.*

Before uncritically using the resulting FFNN, one should compare the quality of its predictions with that obtained using simpler predictors, the simplest possible being the naive predictor

$$\hat{x}(k) = x(k-1). \tag{C22}$$

Scaling back to original units, the error of the best FFNN, i.e. the one with the lowest validation error, is roughly 465 mm, whereas the naive predictor gives a validation error of 595 mm. Thus, even though the prediction error of the FFNN is far from negligible, it does outperform the naive predictor. ∎

Appendix D: Benchmark functions

Clearly, there is an infinitude of functions that could be used as benchmarks in performance investigations. However, the choice is not arbitrary. As a general rule, a benchmark function should preferably contain many *local* optima, making it more difficult to find the *global* optimum (or optima). Alternatively, functions with a single local (and therefore global) optimum can be considered, provided that the approach to the optimum presents some difficulty, for example a very small gradient near the optimum, as is the case for the function Ψ_2 introduced below. Furthermore, the local optima should preferably not be too evenly spaced, since, for instance, some GA components may exploit such even spacing. For example, the function $\Psi_5^{[n]}$ considered below fulfils the criterion of uneven spacing of local optima.

In addition, a benchmark function should not be separable, i.e. it should *not* be possible to write $f(x_1, \ldots, x_n)$ as $\sum_i f_i(x_i)$. Finally, the approach to the global optimum (assuming that there is only one) should not be smooth in any of the coordinate directions. In other words, following a coordinate axis towards the global optimum, one should encounter many local optima along the way. Below, a list of unconstrained optimization problems will be given.[1] With the exception of the function $\Psi_5^{[n]}$ below, the functions are taken from the literature. A few examples of constrained optimization problems can be found in Chapter 6 and a list of many such problems is available in Ref. [21].

---

[1] Strictly speaking, the problems considered here *are*, in fact, constrained, since limits on the variables x_i, of the form $a \leq x_i \leq b$, are introduced. However, at least in the case of GAs, the constraints are trivially fulfilled simply as a result of the encoding procedure, which limits the variable values x_i to a given range. See also Chapter 6.

Benchmark comparisons can be carried out in different ways: either by running the optimization algorithm until a particular value of the objective function (i.e. the fitness value, in the case of a GA) has been reached or by running it until a given number of function evaluations (i.e. individuals, in the case of a GA) have been considered. The former approach is normally much more time-consuming since, in some runs, a very large number of function evaluations may be needed in order to reach the prescribed cutoff value of the objective function.

Regardless of which benchmark function is used, one should keep in mind that a good performance of a given stochastic optimization algorithm on one benchmark function does not in any way guarantee that the same algorithm (with the same parameter settings) will do well on another problem; see the discussion on the NFL theorem in Section 1.2. This is not to say, however, that the choice of algorithm, or its parameters, is arbitrary. For example, as demonstrated in Chapter 6, on most realistic optimization problems, a completely random search (obtained, e.g. by setting the mutation rate p_{mut} to 1 in a GA) does much worse than, say, a GA with a better choice of p_{mut} or PSO with appropriate parameter settings.

D.1 The Goldstein–Price function

The Goldstein–Price function [22] is given by

$$\Psi_1(x_1, x_2) = \left(1 + (x_1 + x_2 + 1)^2 \left(19 - 14x_1 + 3x_1^2 - 14x_2 + 6x_1x_2 + 3x_2^2\right)\right)$$
$$\times \left(30 + (2x_1 - 3x_2)^2 \left(18 - 32x_1 + 12x_1^2 + 48x_2 - 36x_1x_2 + 27x_2^2\right)\right)$$
(D1)

and has a global minimum of 3 at $\mathbf{x}^* = (0, -1)^{\text{T}}$. The typical search range is $-2 \leq x_i \leq 2$, $i = 1, 2$. The function is plotted in the upper left panel of Fig. D.1.

D.2 The Rosenbrock function

This function [64] is given by

$$\Psi_2(x_1, x_2) = 100 \left(x_1^2 - x_2\right)^2 + (1 - x_1)^2,$$
(D2)

and has a global minimum of 0 at $\mathbf{x}^* = (1, 1)^{\text{T}}$. The typical search range is $-5.12 \leq x_i \leq 5.12$, $i = 1, 2$. Even though the function is seemingly simple, the minimum is found at the bottom of a long valley in the search space, and is quite hard to find. The Rosenbrock function can be generalized to n dimensions as

$$\Psi_2^{[n]}(x_1, \ldots, x_n) = \sum_{i=1}^{n-1} \left(100 \left(x_i^2 - x_{i+1}\right)^2 + (1 - x_i)^2\right),$$
(D3)

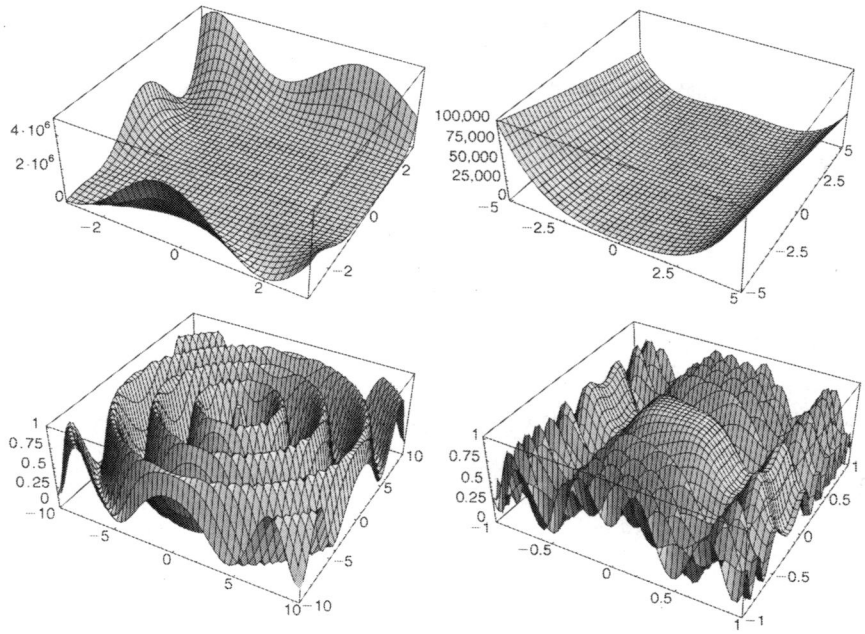

Figure D.1: *Four of the five benchmark functions defined in the text. Top left panel: the Goldstein–Price function; top right panel: the Rosenbrock function, with $n=2$; bottom left panel: the Sine square function; bottom right panel: the benchmark function $\Psi_5^{[2]}$, with $a=0.05$ and $b=10$.*

with a global minimum of 0 at $\mathbf{x}^* = (1,\ldots,1)^T$. The function is plotted for $n=2$ in the upper right panel of Fig. D.1.

D.3 The Sine square function

The Sine square function is one of several functions considered in Ref. [67]. It has the following form

$$\Psi_3(x_1, x_2) = 0.5 - \frac{\left(\sin\sqrt{x_1^2 + x_2^2}\right)^2 - 0.5}{\left(1 + 0.001\left(x_1^2 + x_2^2\right)\right)^2}, \tag{D4}$$

with a global maximum of 1 at $\mathbf{x}^* = (0,0)^T$. The typical search range is $-100 \leq x_i \leq 100$, $i=1,2$. The function is plotted in the lower left panel of Fig. D.1.

D.4 The Colville function

The Colville function [10] is defined as

$$\Psi_4(x_1,x_2,x_3,x_4) = 100\left(x_2 - x_1^2\right)^2 + (1-x_1)^2 + 90\left(x_4 - x_3^2\right)^2 + (1-x_3)^2$$
$$+ 10.1((x_2 - 1)^2 + (x_4 - 1)^2) + 19.8(x_2 - 1)(x_4 - 1). \tag{D5}$$

The search range is $-10 \leq x_i \leq 10$, $i = 1, \ldots, 4$, and the global minimum $\Psi_4^* = 0$ occurs at $\mathbf{x}^* = (1, 1, 1, 1)^{\mathrm{T}}$.

D.5 A multidimensional benchmark function

This function, which bears no particular name, is defined as

$$\Psi_5^{[n]}(x_1,x_2,\ldots,x_n) = \frac{1}{2} + \frac{1}{2n} \exp\left(-a \sum_{i=1}^n x_i^2\right) \sum_{i=1}^n \cos\left(b\sqrt{i}x_i \sum_{j=1}^i jx_j\right), \tag{D6}$$

where a and b are two positive parameters. The global maximum value of 1 is attained at $\mathbf{x}^* = (0, \ldots, 0)^{\mathrm{T}}$. The typical search range is $-3 \leq x_i \leq 3$, $i = 1, \ldots, n$. The function is plotted for $n = 2$, $a = 0.05$ and $b = 10$ in the lower right panel of Fig. D.1.

Answers to selected exercises

Chapter 2

2.1 The stationary points occur at $(x_1,x_2)^T = (0,0)^T$, $(-5/3, 0)^T$, $(-1, 2)^T$ and $(-1, -2)^T$.

2.2 $f(x_1, x_2)$ has a local minimum (equal to 0) at $(x_1, x_2)^T = (0, 0)^T$.

2.4 The composition $g(h(x))$ is convex if both $g(z)$ and $h(x)$ are convex, *and* $g(z)$ is a non-decreasing function, i.e. $g'(z) \geq 0 \ \forall z$.

2.7 The gradient of $f(x_1, x_2)$ at $(1, 1)^T$ equals $(5, 3)^T$. Line search using, for example, the bisection method, gives $\eta^* \approx 0.2722$, and the point reached after one step of gradient descent is $(-0.3606, 0.1836)^T$. At this point, the gradient equals $(-0.0040, 0.0066)^T$, a vector that is orthogonal to the gradient at $(1, 1)^T$.

2.10 The minimum value f^* equals -2.

2.11 The minimum value f^* equals -10.

2.13 The maximum value f^* equals 24.

Chapter 3

3.2 (1) Roulette-wheel selection gives $p_4 = \frac{2}{9} \approx 0.222$. (2) Tournament selection gives $p_4 = \frac{6}{25} = 0.240$. (Assuming that the random choice of the two individuals included in the tournament is done with replacement). (3) Roulette-wheel selection with fitness ranking gives $p_4 \approx 0.282$.

3.6 $d(S_1) = 4$, $o(S_1) = 2$, $\Gamma(S_1, g) = 2$, $\bar{f}(S_1) = 0.80859$, $\bar{f} = 0.4954 \Rightarrow \Gamma(S_1, g+1) \approx 3$.

3.7 (a) $p(\text{no mutations}) = (1-p)^m$, (b) $p(\text{one mutation}) = mp(1-p)^{m-1}$, (c) $p(\text{less than three mutations}) = (1-p)^m + mp(1-p)^m + [(m(m-1))/2]p^2(1-p)^{m-2}$.

3.10 (a) 10, (b) 35, (c) $\binom{n+d}{n} \equiv \binom{n+d}{d}$.

3.11 (a) $P^+(m,j,k,p) = \sum_{s=0}^{\min(j,m-j-k)} \binom{m-j}{s+k}\binom{j}{m}p^{k+2s}(1-p)^{m-k-2s}$.

(b) $P^e(10,1,0.1) = 0.5738$, $P^e(10,5,0.1) = 0.2692$, $P^e(10,9,0.1) = 0.0387$. The results obtained with the approximate expression are 0.5513, 0.2418 and 0.0387, respectively. Hence, it can be concluded that the approximate expression gives rather good results.

Chapter 4

4.3 The shortest paths have lengths 90.01 ($n=30$), 110.17 ($n=50$) and 155.65 ($n=100$) length units, respectively.

Chapter 5

5.2 The minimum value $f^* = 1$ is obtained for $(x_1, x_2)^T = (5, 4)^T$.

5.6 The minimum value $f^* = -6$ is obtained for $(x_1, x_2)^T = (2, -1)^T, (3, -2)^T, (3, -1)^T$ and $(4, -2)^T$.

Bibliography

[1] N. Andréasson, A. Evgrafov, and M. Patriksson. *An Introduction to Continuous Optimization*. Studentlitteratur, 2005.
[2] P. Ball. Shark skin and other solutions. *Nature*, 400:507–508, 1999.
[3] A. Bauer, B. Bullnheimer, R. F. Hartl, and C. Strauss. Minimizing total tardiness on a single machine using ant colony optimization. *Central European Journal for Operations Research and Economics*, 8:125–141, 2000.
[4] J. C. Bongard and R. Pfeifer. Evolving complete agents using artificial ontogeny. In: *Morpho-Functional Machines: The New Species*. Springer Verlag, pp. 237–258, 2003.
[5] S. Boyd and L. Vandenberghe. *Convex Optimization*. Cambridge University Press, 2004.
[6] M. Brameier. *On Linear Genetic Programming*. PhD thesis, Dortmund University, 2003.
[7] M. Byröd. *Theoretical Aspects of Genetic Algorithms*. M.Sc. thesis, Chalmers University of Technology, 2005.
[8] M. Clerc. The swarm and the queen: Toward a deterministic and adaptive particle swarm optimization. In: *Proceedings of the IEEE International Conference on Evolutionary Computation*, pp. 1951–1957, 1999.
[9] M. Clerc and J. Kennedy. The particle swarm: Explosion, stability, and convergence in a multi-dimensional complex space. *IEEE Transactions on Evolutionary Computation*, 6:58–73, 2002.
[10] A. R. Colville. A comparative study on nonlinear programming codes. *IBM Scientific Center Report 320-2949*, 1968.
[11] R. Dawkins. *The Blind Watchmaker*. Norton, 1986.
[12] R. Dawkins. *Climbing Mount Improbable*. Penguin, 1996.
[13] K. Deb and A. R. Reddy. Reliable classification of two-class cancer data using evolutionary algorithms. *Biosystems*, 72:111–129, 2003.
[14] M. L. den Besten, T. Stützle, and M. Dorigo. Ant colony optimization for the total weighted tardiness problem. *Lecture Notes in Computer Science*, 1917:611–620, 2000.
[15] J.-L. Deneubourg, S. Aron, S. Goss, and J.-L. Pasteels. The self-organizing exploratory pattern of the argentine ant. *Journal of Insect Behavior*, 3:159–168, 1990.
[16] J. M. Deutsch. Evolutionary algorithms for finding optimal gene sets in microarray prediction. *Bioinformatics*, 19:45–52, 2003.
[17] M. Dorigo. *Optimization, Learning and Natural Algorithms [in Italian]*. PhD thesis, Dipartimento di Elettronica, Politecnico di Milano, 1992.
[18] M. Dorigo, V. Maniezzo, and A. Colorni. Ant system: Optimization by a colony of cooperating agents. *IEEE Transactions on Systems, Man, and Cybernetics, Part B*, 26:29–41, 1996.
[19] P. Eggenberger. Creation of neural networks based on developmental and evolutionary principles. In: *Proceedings of the International Conference on Artificial Neural Networks*, pp. 337–342, 1997.

[20] R. A. Fisher. Applications of "Student's" distribution. *Metron*, 5:90–104, 1925.
[21] C. A. Floudas and P. M. Pardalos. *A Collection of Test Problems for Constrained Global Optimization Problems*, volume 455 of *Lecture Notes in Computer Science*. Springer Verlag, 1990.
[22] A. A. Goldstein and J. F. Price. On descent from local minima. *Mathematics of Computation*, 25:569–574, 1971.
[23] B. Gompertz. On the nature of the function expressive of the law of human mortality, and on a new mode of determining the value of life contingencies. *Philosophical Transactions of the Royal Society of London*, 115:513–585, 1825.
[24] W. S. Gosset. The probable error of a mean. *Biometrika*, 6:1–25, 1908.
[25] F. Gray. Pulse code communication, 1953. U.S. Patent 2,632,058.
[26] V. G. Gudise and G. K. Venayagamoorthy. Comparison of particle swarm optimization and backpropagation as training algorithms for neural networks. In: *Proceedings of the IEEE Swarm Intelligence Symposium*, pp. 110–117, 2003.
[27] W. J. Gutjahr. First steps to the runtime complexity analysis of ant colony optimization. *Computers and Operations Research*, 35: 2711–2727, 2008.
[28] S. Haykin. *Neural Networks: A Comprehensive Foundation*. Prentice Hall, 1994.
[29] D. O. Hebb. *The Organization of Behavior*. Wiley, 1949.
[30] D. Hillis. Co-evolving parasites improve simulated evolution in an optimization procedure. *Physica D*, 42:228–234, 1990.
[31] K. W. Hipel and A. I. McLeod. *Time Series Modelling of Water Resources and Environmental Systems*. Elsevier, 1994.
[32] J. H. Holland. *Adaptation in Natural and Artificial Systems*. University of Michigan Press, 1975.
[33] J. J. Hopfield. Neural networks and physical systems with emergent collective computational abilities. *Proceedings of the National Academy of Sciences of the USA*, 79:2554–2558, 1982.
[34] E. R. Kandel. The molecular biology of memory storage: A dialog between genes and synapses. *Bioscience Reports*, 24:475–522, 2004.
[35] E. R. Kandel, J. H. Schwartz, and T. M. Jessell. *Principles of Neural Science*. McGraw-Hill, 4th edition, 2000.
[36] J. Kennedy and R. C. Eberhart. Particle swarm optimization. In *Proceedings of the IEEE International Conference on Neural Networks*, pp. 1942–1948, 1995.
[37] J. Kennedy and R. C. Eberhart. A discrete binary version of the particle swarm algorithm. In: *Proceedings of the 1997 Conference on Systems, Man, and Cybernetics*, pp. 4104–4109, 1997.
[38] T. Kohonen. *Self-Organizing Maps*. Springer Verlag, 3rd edition, 2001.
[39] J. R. Koza. *Genetic Programming: On the Programming of Computers by Means of Natural Selection*. MIT Press, 1992.
[40] C. R. Kube and H. Zhang. Collective robotics: From social insects to robots. *Adaptive Behavior*, 2:189–218, 1994.
[41] W. B. Langdon and R. Poli. *Foundations of Genetic Programming*. Springer Verlag, 2003.
[42] W. Latham and S. Todd. *Evolutionary Art and Computers*. Academic Press, 1992.
[43] B. Lewin. *Genes VIII*. Prentice-Hall, 2004.
[44] L. Li, L. G. Pedersen, T. A. Darden, and C. Weinberg. Computational analysis of leukemia microarray expression data using the GA/KNN method. In: *Proceedings of the First Conference on Critical Assessment of Microarray Data Analysis*, 2000.

[45] P. Lingman and M. Wahde. Transport and maintenance effective retardation control using neural networks with genetic algorithms. *Vehicle System Dynamics*, 42:89–107, 2004.

[46] W. Z. Lu, H. Y. Fan, A. Y. T. Leung, and J. C. K. Wong. Analysis of pollutant levels in central Hong Kong applying neural network method with particle swarm optimization. *Environmental Monitoring and Assessment*, 79:217–230, 2002.

[47] International Human Genome Sequencing Consortium. Finishing the euchromatic sequence of the human genome. *Nature*, 431:931–945, 2004.

[48] W. McCulloch and W. Pitts. A logical calculus of the ideas immanent in nervous activity. *Bulletin of Mathematical Biophysics*, 7:115–133, 1943.

[49] Z. Michalewicz and M. Schoenauer. Evolutionary algorithms for constrained parameter optimization problems. *Evolutionary Computation*, 4:1–32, 1996.

[50] M. Moffett. Cooperative food transport by an Asiatic ant. *National Geographic Research*, 4:386–394, 1988.

[51] F. Mondada, G. Pettinaro, A. Guignard, I. Kwee, D. Floreano, J.-L. Deneubourg, S. Nolfi, L. M. Gambardella, and M. Dorigo. Swarm-bot: A new distributed robotic concept. *Autonomous Robots*, 17: 193–221, 2004.

[52] H. Mühlenbein. How genetic algorithms really work: Mutation and hill-climbing. In: *Parallel Problem Solving From Nature*, pp. 15–26, 1992.

[53] S. Nolfi and D. Floreano. *Evolutionary Robotics: The Biology, Intelligence, and Technology of Self-Organizing Machines*. MIT Press, 2000.

[54] E. Oczan and C. K. Mohan. Analysis of a simple particle swarm optimization system. In: *Intelligent Engineering Systems through Artificial Neural Networks*, pp. 253–258, 1998.

[55] P. Palangpour, G. K. Venayagamoorthy, and K. Duffy. Recurrent neural network based predictions of elephant migration in a South African game reserve. In: *Proceedings of the 2006 International Joint Conference on Neural Networks*, pp. 4084–4088, 2006.

[56] I. R. Pavlov. *Conditioned Reflexes: An Investigation of the Physiological Activity of the Cerebral Cortex*. Oxford University Press, 1927. Translated by G.V. Anrep.

[57] A. Petrovski, B. Sudha, and J. McCall. Optimising cancer chemotherapy using particle swarm optimization and genetic algorithms. In: *Parallel Problem Solving From Nature*, pp. 633–641, 2004.

[58] R. Poli. An analysis of publications on particle swarm optimisation applications. *Technical Report CSM-469, Department of Computer Science, University of Essex*, 2007.

[59] M. Ptashne and A. Gann. *Genes and Signals*. Cold Spring Harbor Laboratory Press, 2002.

[60] S. S. Rao. *Engineering Optimization – Theory and Practice*. Wiley Eastern, 1996.

[61] C. R. Reeves and J. E. Rowe. *Genetic Algorithms – Principles and Perspectives*. Kluwer Academic Publishers, 2003.

[62] C. W. Reynolds. Flocks, herds, and schools: A distributed behavioral model. *Computer Graphics*, 21:25–34, 1987.

[63] J. Romero and P. Machado, editors. *The Art of Artificial Evolution: A Handbook on Evolutionary Art and Music*. Springer Verlag, 2007.

[64] H. H. Rosenbrock. An automatic method for finding the greatest or least value of a function. *Computer Journal*, 3:175–184, 1960.

[65] A. H. Rots, A. Bosma, J. M. van der Hulst, E. Athanassoula, and P. C. Crane. High-resolution HI observations of the whirlpool galaxy M51. *Astronomical Journal*, 100:387–393, 1990.

[66] S. S. Sawilowsky. Fermat, Schubert, Einstein, and Behrens-Fisher: The probable difference between two means when $\sigma_1^2 \neq \sigma_2^2$. *Journal of Modern Applied Statistical Methods*, 1:461–472, 2002.
[67] J. D. Schaffer, R. Caruana, L. J. Eshelman, and R. Das. A study of control parameters affecting the online performance of genetic algorithms for function optimization. In: *Proceedings of the 3rd International Conference on Genetic Algorithms*, pp. 51–60, 1989.
[68] R. O. Schoonderwoerd, O. Holland, J. Bruten, and L. Rothkrantz. Ant-based load balancing in telecommunications networks. *Adaptive Behavior*, 5:169–207, 1996.
[69] M. Settles and T. Soule. Breeding swarms: A GA/PSO hybrid. In: *Proceedings of the 2005 conference on Genetic and Evolutionary Computation (GECCO2005)*, pp. 161–168, 2005.
[70] T. Stützle and H. H. Hoos. Max–min ant system. *Future Generation Computer Systems*, 16:889–914, 2000.
[71] H. Takagi. Interactive evolutionary computation: Fusion of the capabilities of EC optimization and human evaluation. *Proceedings of the IEEE*, 89:1275–1296, 2001.
[72] L. J. vant Veer, H. Dai, M. J. van de Vijver, Y. D. He, A. A. Hart, M. Mao, H. L. Peterse, K. van der Kooy, M. J. Marton, A. T. Witteveen, G. J. Schreiber, R. M. Kerkhoven, C. Roberts, P. S. Linsley, R. Bernards, and S. H. Friend. Gene expression profiling predicts clinical outcome of breast cancer. *Nature*, 415:484–485, 2002.
[73] M. D. Vose. *The Simple Genetic Algorithm*. MIT Press, 1990.
[74] M. Wahde. Determination of orbital parameters of interacting galaxies using a genetic algorithm. Description of the method and application to artificial data. *Astronomy and Astrophysics, Supplement Series*, 132:417–429, 1998.
[75] M. Wahde. *An Introduction to Adaptive Algorithms and Intelligent Machines*. Biblioteks Reproservice, Chalmers University of Technology, 5th edition, 2006.
[76] M. Wahde and K. J. Donner. Determination of the orbital parameters of the M51 system using a genetic algorithm. *Astronomy and Astrophysics*, 379:115–124, 2001.
[77] M. Wahde and Z. Szallasi. Classification of gene expression data using evolutionary algorithms. *Expert Review of Molecular Diagnostics*, 6:101–110, 2006.
[78] M. Wahde and Z. Szallasi. Improving the prediction of the clinical outcome of breast cancer using evolutionary algorithms. *Soft Computing*, 10:338–345, 2006.
[79] B. Widrow and M. E. Hoff. Adaptive switching circuits. *IRE WESCON Convention Record*, pp. 96–104, 1960.
[80] D. H. Wolpert and G. W. Macready. No free lunch theorems for optimization. *IEEE Transactions on Evolutionary Computation*, 1:67–82, 1997.
[81] L. Wolsey. *Integer Programming*. Wiley, 1998.

Index

n-body simulations, 87
n-parity function, 169

activation function, 159
activity, 82
adaptation, 2
affine function, 24
allele, 39
amino acid, 37
ant colony optimization, 100
Ant-colony system, 113
Ant system, 105
asexual reproduction, 39
axon, 151

backpropagation, 161
backpropagation, recurrent, 171
base, 36
batch training, 162
behaviour selection, 114
behaviour-based robotics, 114
bias term, 158
binary classification, 94
binary encoding scheme, 40
binary integer programming, 12
binomial distribution, 197
Biomorph, 79
bisection method, 17
Box-Müller transform, 196
branch-and-bound method, 113
breeding swarms, 140
building blocks, 60

central limit theorem, 195
chromosome, 36
chromosome (GA), 40
class discovery, 94
clustering algorithm, 94

codon, 37
co-evolution, 3
cognitive component, 123
compact, 29
confidence interval, 197
conjugate gradient, 20
connection weight, 156
constraint function, 11
constraint, equality, 11
constraint, inequality, 11
constriction coefficient, 127
construction graph, 104
construction graph, chain, 104
convex function, 14
convex optimization problems, 24
convex set, 14
craziness operator, 128
credit assignment problem, 162
creep rate, 54
critical point, 12
crossover (GA), 42
crossover averaging, 53
crossover point (GA), 42
crossover probability, 53
crossover, length-preserving, 53
crossover, single-point, 53
crossover, uniform, 53

Darwin's theory of evolution, 36
data classification, 94
decision variable, 10
defining length, 60
Delta rule, 161
dendrite, 151
destination register, 74
diffusion model, 70
diploid, 36
double-stranded, 36

earliest due date, 113
elitism, 55
Elman network, 170
encoding scheme, 40
enzyme, 37
epoch, 165
Euclidean norm, xv
evaporation rate, 106
evolutionary algorithm, 35
evolutionary art, 79
evolutionary music, 80
exon, 82
expectation value, 194
extremum, local, 9

feasible point, 11
fitness, 39
fitness (GA), 41
fitness ranking, 51
Frank-Wolfe algorithm, 25
function of unitation, 63

gene, 36
gene (GA), 40
gene, regulatory, 39
gene, structural, 39
generation (GA), 40
generational gap, 55
genetic algorithm, 40
genetic programming, 72
genetic regulatory network, 82
genome, 36
genotype, 39
germ cell, 36
gradient, 13
gradient descent, 19
gradient descent, stochastic, 162
Gray code, 47

habituation, 155
Hamming distance, 62
haploid, 36
Hessian, 14
hidden layer, 157
holdout validation, 201
Hopfield network, 172
hypothesis evaluation, 193

if-then-else rule, 5
inertia weight, 128
infimum, 13
input element, 157
input-output pair, 160
insertion (ACO local search), 113
instruction (LGP), 73
instruction set, 73
integer programming, 12
interactive evolutionary computation, 78
interchange (ACO local search), 113
interior point method, 25
interval halving, 18
intron, 82
island model, 70
iteration (ACO), 106
iteration (PSO), 124

job shop scheduling, 112
Jordan network, 170

Karush-Kuhn-Tucker conditions, 28
Kohonen map, 171

Lagrange multiplier, 26
Lagrange multiplier method, 26
learning method, supervised, 157
learning method, unsupervised, 157
learning, strongly guided, 157
learning, weakly guided, 157
Lewenberg-Marquardt algorithm, 24
line search, 17
linear programming, 11
linear genetic programming (LGP), 73
local gradient, 163
local search, 113
logistic sigmoid, 159
long-term depression, LTD, 154
long-term potentiation, LTP, 154

macromutation, 69
Markov process, 61
mass-to-light ratio, 89
mathematical programming, 1
mating restriction, 70
maximum, global, 10
maximum, local, 9
Max-min ant system, 109

McCulloch-Pitts (MCP) neuron, 158
messenger RNA, 37
messy encoding, 47
microarray, 92
minimum, global, 10
minimum, local, 9
multicellularity, 83
modified due date, 113
mutation, 39
mutation probability (GA), 43
mutation, parametric, 96
mutation, structural, 96

nearest-neighbour tour, 106
negative definite, 14
neighbourhood, 9
neighbourhood (PSO), 125
neural network, artificial, 151
neural network, feedforward, 156
neural network, recurrent, 156
Newton's modified method, 22
Newton-Rhapson's method, 22
No-free-lunch theorem, 4
non-linear programming, 12
normal distribution, 195
null hypothesis, 197

objective function, 10
offspring, 53
Onemax, 63
operand (LGP), 74
operator (biology), 82
operator (GP), 72
operator (LGP), 74
optimization, 1
optimization problem, constrained, 11
optimization problem, continuous, 11
optimization problem, differentiable, 11
optimization problem, unconstrained, 11
optimization, classical, 9
optimization, combinatorial, 12
optimization, deterministic, 2
optimization, stochastic, 2
optimum, local, 9
order, 60
order crossover, 147
output layer, 157
overfitting, 201

p-value, 198
parent, 53
particle swarm optimization, 118
penalty function, 31
penalty method, 31
periodic boundary conditions, 71
permutation encoding, 48
phenotype, 39
pheromone, 100
physical graph, 104
population, 36
population (GA), 40
positive definite, 14
positive semi-definite, 14
premature convergence, 68
probability distribution, 194
promoter region, 82
protected definition, 74
protein, 37

quadratic programming, 12
quasi-Newton methods, 24

radial-basis function network, 171
raw fitness value, 51
real-number creep, 54
recruitment, 99
recurrent neural network,
    continuous-time, 170
refractory period, 152
register, 73
register, constant, 73
register, variable, 73
relative range of due dates, 113
replacement, 43
replacement, generational, 43
replacement, steady-state, 55
reproduction, 39
reproduction (GA), 42
restricted three-body simulation, 89
ribosome, 37
RMS error, 78
RNA polymerase, 37
run, xvi

s-Bot, 115
saddle point, 13

sampling distribution of the mean, 195
scheduling problem, 112
Schema theorem, 59
search space, 40
selection (GA), 42
selection pressure, 51
selection, Boltzmann, 51
selection, roulette-wheel, 48
selection, tournament, 48
self-gravitating simulation, 89
self-organizing feature map
    (SOFM), 172
sensitization, 155
sexual reproduction, 39
shortest processing time, 112
short-term potentiation, STP, 154
sigmoid function, 159
Simplex method, 25
single-machine total weighted tardiness
    problem, 112
social component, 123
somatic cell, 36
species, 36
spike, 152
squashing function, 159
standard deviation, 195
standard error, 195
stationary point, 12
stigmergy, 101

stochastic programming, 2
strictly convex, 15
Student's t-distribution, 199
subgradient method, 25
swap mutation, 147
Swarm-bots, 115
synapse, 151

tabu list, 105
tardiness factor, 113
terminal, 72
test set, 201
tournament selection parameter, 50
tournament size, 50
training element, 160
training set, 201
transcription, 37
transfer RNA, 37
translation, 37
transmitter substance, 152
travelling salesman problem, 104

validation set, 201
variance, 194
visibility, 106

weight, 156
Widrow–Hoff rule, 161

 **WIT**PRESS ...for scientists by scientists

Design & Nature IV
Comparing Design in Nature with Science and Engineering

Edited by: **C.A. BREBBIA**, *Wessex Institute of Technology, UK*

Throughout history, the parallels between nature and human design, in mathematics, engineering and other areas, have inspired many leading thinkers. Today, the huge increase in biological knowledge, developments in design engineering systems, together with the virtual revolution in computer power and simulation modelling, have all made possible more comprehensive studies of nature.

Scientists and engineers now have at their disposal a vast array of relationships for materials, mechanisms and control. The resulting laws have been painstakingly assembled by observation and analysis and span the cosmic scale of space down to the molecular level of genetics. In particular, they have made us aware of the rich diversity of the natural world around us.

These developments are presented here in the proceedings of the Fourth International Conference on Comparing Design in Nature with Science and Engineering. The Conference topics include: Shape and Form in Engineering and Nature; Nature and Architectural Design; Thermodynamics in Nature; Biomimetics; Natural Materials in Engineering; Mechanics in Nature; Bioengineering; Solution from Nature; Complexity; Sustainability Studies; Education and Training.

This book should be of interest to researchers and those interested in the study of natural materials, organisms, processes and their significance for design in the world today.

WIT Transactions on Ecology and the Environment, Vol 114
ISBN: 978-1-84564-120-7   2008   368pp   £121.00/US$242.00/€181.50

WITPress
Ashurst Lodge, Ashurst,
Southampton,
SO40 7AA, UK.
Tel: 44 (0) 238 029 3223
Fax: 44 (0) 238 029 2853
E-Mail: witpress@witpress.com

 ...for scientists by scientists

Design and Information in Biology

From Molecules to Systems – Volume 2

Edited by: **J.A. BRYANT**, *University of Exeter, UK,* **M. ATHERTON**, *Brunel University West London UK and* **M.W. COLLINS**, *Brunel University West London, UK*

Highlighted with individual contributions from eminent specialists, these multi authored volumes combine authority, inspiration and state-of-the-art knowledge. Both informative and inspiring they are designed to appeal to scientists and interested people alike.

Volume 2 complements and extends the scope of the first, with the biological viewpoint being stressed. Following an introductory chapter on design as understood in biology, the various aspects of the biological information revolution are addressed. Areas discussed include molecular structure, the genome, development, and neural networks. A section on information theory provides a link with engineering, and the scope is also broadened to include the implications of motion in nature and engineering.

Series: Design & Nature, Vol 2
ISBN: 1-85312-853-8   2007   512pp   £165.00/US$285.00/€247.50

WIT eLibrary

Home of the Transactions of the Wessex Institute, the WIT electronic-library provides the international scientific community with immediate and permanent access to individual papers presented at WIT conferences. Visitors to the WIT eLibrary can freely browse and search abstracts of all papers in the collection before progressing to download their full text.

Visit the WIT eLibrary at
http://library.witpress.com